W9-CLE-389

Public Policy Issues in Resource Management

James A. Crutchfield and Robert H. Pealy, Editors

Public Policy Issues in Resource Management

1. The Fisheries: Problems in Resource Management, edited by James A. Crutchfield
2. Water Resources Management and Public Policy, edited by Thomas H. Campbell and Robert O. Sylvester
3. Weather Modification: Science and Public Policy, edited by Robert G. Fleagle
4. World Fisheries Policy: Multidisciplinary Views, edited by Brian J. Rothschild
5. Ocean Resources and Public Policy, edited by T. Saunders English

World Fisheries Policy

Multidisciplinary Views

Edited by Brian J. Rothschild

University of Washington Press

Seattle and London

Printed in the United States of America

Library of Congress Cataloging in Publication Data

Main entry under title:

World fisheries policy.

(Public policy issues in resource management, v. 4)
Based on papers presented at a series of seminars sponsored by the Graduate School of Public Affairs, University of Washington.
Includes bibliographies.
1. Fishery policy. I. Rothschild, Brian J., 1934– ed. II. Washington (State). University. Graduate School of Public Affairs. III. Series.
SH323.W67 338.3'72'7 72-8927
ISBN 0-295-95232-6

Dedication

Wilbert McLeod Chapman, 1910–70
Milner Baily Schaefer, 1912–70

Turn safe to rest, no dreams, no waking;
And here, man, here's a wreath I've made;
'Tis not a gift that's worth the taking,
But wear it and it will not fade.

A. E. Housman

Two lifelong, staunch, inseparable, and highly argumentative friends. Two dedicated scientists and fearless searchers for truth. Two independent, different, but brilliant and parallel careers cut short.

Born within two years of each other, they met and struggled as undergraduates at the University of Washington's School of Fisheries, establishing solid academic records. Both graduated into the depths of a severe economic depression when the only "professional" job available anywhere was sorting marine plankton for W. F. Thompson of the International Pacific Halibut Commission.

Both volunteered for military service during World War II. Chapman, because of a slight defect, was turned down. He promptly applied to and was accepted by the Board of Economic Warfare to go fishing for the troops in the war zone of the South Pacific. This experience supplied material for his timely book, *Fishing in Troubled Waters.** The two friends still managed to serve together in the same theater of war, one in the Navy and the other catching fish for the Army.

They studied together toward their Ph.D.'s and they gained experience and made their independent reputations and garnered fame in their chosen field, the scientific and philosophical study of fisheries, fueling and firing each other on their way.

They lived the last decades of their lives and brought up their colorful and substantial families near each other on a lovely San Diego hill. And

* Philadelphia: J. B. Lippincott, 1949.

they died within a month of each other, at the peak of their worldwide activities and their professional careers. With their passing something dynamic, original, imaginative, and colorful was suddenly lost. And there is nothing in sight anywhere in the world in the fisheries field that even remotely promises to fill the void.

Although there were many similarities between these two great friends the differences were even greater.

Wilbert McLeod Chapman, known universally as "Wib," was a great, gruff bear of a man, with a resonant bass voice and a glowing personality. His thunderous laugh could fill any room. He was low keyed and calm and not easily ruffled, yet his presence was felt in any company.

Wib was born on March 31, 1910, in Kalama, Washington, the son of parents of Scottish descent. He received all his formal education in the state of Washington from grammar school through the Ph.D. degree which he received in 1937 from the University of Washington. In 1935 he married Mary Elizabeth (Mazie) Swaney, a fellow student at the University. This happy union resulted in six children, four boys and two girls.

The formal positions held by Wib Chapman, though many, are easy to list, but his extracurricular professional and other (including business, diplomatic, and political) activities were so numerous and so varied that these can only be sampled.

On graduation the first professional job he held, as already noted, was with the International Pacific Halibut Commission. As his personal responsibilities increased, he was constantly on the lookout for a position that carried more pay. This led him to accept a post in 1935 as assistant biologist with the Washington State Department of Fisheries. This was in many ways a fortunate move as it gave him a broad experience. He studied at one time or another the life histories and dynamics of salmon, herring, sardines, oysters, and clams. In 1941 he moved from Washington state to work on fur seal research for the United States Fish and Wildlife Service.

But Wib's first stated love in science was in the field of systematic ichthyology. Even as an undergraduate he started a study which he immodestly called "A Systematics Study of the Blennies of the World." He wanted a position where he could continue this and other similar studies. In late 1943 he found such an opportunity, when he was offered and immediately accepted a position as curator of fishes at the California Academy of Sciences. It was from this position that he took leave of absence to catch fish for the Army in the South Pacific.

This happy situation was not to last long. In 1947, just ten years after he had completed his own studies at the University of Washington School of Fisheries, he was named director of the school and professor of fisheries. For most people this would be the end of the "local boy makes good" story, but for Wib it was just the beginning. He never functioned at his best in a fixed routine. He was a free-wheeling individual who needed much room to maneuver. This he missed at the University. So when, in response to much displeasure within the fishing industry concerning the government's handling of international fisheries matters, the Department of State in 1948 went scouting for a man qualified to head a fisheries unit in the department. Wib applied for, and got the job. The position carried the title of Special Assistant for Fish and Wildlife to the Under-Secretary of State. This was a new job in an old department and the size and activities of the position had yet to be established.

Wib's tempo for getting things done and that of the career diplomats soon came in conflict. When he steered the Inter-American Tropical Tuna Treaty through signing by the governments and ratification by the Senate in three weeks instead of months or years as was the custom, the establishment was horrified. At that he managed to stay with the department until 1951. His greatest contribution here was to impress on the senior officers of State that high seas fisheries, and oceanography in general, was very much a problem of international diplomacy and bargaining, that the subject was growing in importance, and that people knowledgeable in this field were needed in the department to further United States and world interests. This unit in the Department of State has been growing in size and rank and perhaps even in importance ever since.

But working within the confines of government once more proved too restrictive for this energetic frame and this active and original mind. In 1951 when offered a job with the title of director of research by the American Tunaboat Association, which was based in San Diego and for which he was invited to write his own job description, he accepted immediately. From this day on his freedom of movement and his choice of problem seemed to be entirely unrestricted. The broad description he gave to his present and subsequent duties was "the application of science and technology to fishery development."

He stayed with the American Tunaboat Association until 1959. Then during 1959–60 he was director, Resources Committee. At this time he established his headquarters and office in his own home overlooking beautiful San Diego Bay. Later in 1961 he became an executive with Van Camp Seafood Company and later with Ralston Purina Company when the latter bought out Van Camp. Though his titles and his employers changed, his office in his home remained his headquarters, and his activities were in no way restricted. He was serving in this capacity when he died.

The above were his formal jobs. Just how he served his employers was never very clear since his extracurricular professional and other activities took up so much of his time and energy. But his employers always seemed eminently satisfied.

Among a legion of other activities he served on committees for the Food and Agriculture Organization of the United Nations and led the fight for the reorganization and upgrading of the fisheries unit of that organization. He served on scientific committees of the National Academy of Sciences, the Coast Guard, the California Department of Fish and Game, the Department of the Interior, the Law of the Sea Institute–University of Rhode Island, the Department of State, the Sea Grant College Program of the National Science Foundation, and many others. During the last decade of his life there was hardly a meeting held anywhere in the world on fisheries or on oceanography in general that he did not attend, and the outcome of which he did not influence. And for his efforts he was elected "Fisheries Man of the Year" in 1966 by the National Fisheries Institute and he earned the First Sea Grant College Award in 1968.

With all this activity he still found time to serve as expert witness before numerous congressional committees, to author some 250 papers on every aspect of fisheries imaginable, and to write letters (often up to thirty or forty typewritten pages) to his friends and associates. These latter were so frequent and so voluminous that laughingly his friends would say they measured their length by the pound. If the letters were marked "personal," the saying was, Wib would send copies to only three hundred individuals. If marked "confidential," however, only one hundred would receive copies. In these letters he would cover any subject or problem he was trying to influence at the time. All letters were interesting and all contained some humor no matter how serious or important the content. One collector of Wib's letters proudly states that he has nearly one hundred pounds in storage.

At the time of his death Wib was planning to attend and actively participate at another meeting scheduled to be held in Malta on ocean uses and ownership. This time he was going to take his wife and members of his family. The trip was never made. But it would have been a more useful and vivacious meeting had he been there.

Milner Baily Schaefer, known to his friends as "Benny," by way of contrast was a slight, somewhat intense man with occasional nervous mannerisms. Though a genial host and a convivial companion he was not as outward going as his friend. His chief interest in life was science and scientific research and he contributed mightily in his chosen field during the whole of his active lifetime.

Schaefer was born in Cheyenne, Wyoming, on December 14, 1912, to German-American parents. He was a brilliant student who graduated from the University of Washington, School of Fisheries, in 1935 Magna Cum Laude and won the President's Medal for scholastic excellence. He received his Ph.D. degree, which had been interrupted by the war, in 1950. In 1949 he married Isabella Long whom he met while on a temporary assignment in Washington, D.C. They had three children, two boys and a girl.

As in Chapman's case Schaefer's formal positions are easy to enumerate but his extracurricular professional activities were legion. He had great competence but little real interest in diplomacy or politics. His abiding interest in his own words was "finding out how the world is put together." His scientific integrity was unquestionable.

Schaefer's first professional job, like Chapman's, was sorting marine plankton for W. F. Thompson, his principal teacher as well as his employer. Because of his restless nature and active mind he moved from this rather mechanical activity as soon as opportunity allowed.

From 1935 to 1939 he worked as assistant biologist for the Washington State Department of Fisheries, where he applied his incisive mind on a variety of problems. From there he moved to work for the International Pacific Salmon Fisheries Commission again under W. F. Thompson's direction. It was from this position that he left in 1942 to serve as line officer in the United States Navy. He terminated his naval assignment in 1946. While in military service he contracted rheumatic fever which was to plague him off and on for the rest of his life.

On returning to civilian life he served for a short period as instructor at the School of Fisheries, University of Washington. From there he went (1946) to work for the South Pacific Fishery Investigations of the United States Fish and Wildlife Service, headquartered at Stanford University. When the Pacific Oceanic Fishery Investigations were initiated in 1948 by the USFWS in Honolulu under special congressional act, he transferred to the position of chief of research and development for this central Pacific activity.

The Inter-American Tropical Tuna Treaty was quickly ratified in 1950 with help from the new special assistant in State. A commission was established to carry out the dictates of the new convention, which were "to study the biology, ecology and population dynamics of the tunas . . . of the eastern Pacific . . . and to recommend conservation measures . . . when researches show such measures to be necessary." Schaefer was unanimously elected to serve as a commission's first director of investigations. Here was a challenge and an opportunity not only to study and recommend, but to implement the recommendations.

On the basis of studies conducted by Schaefer and a group of bright young men and women he gathered around him, he was able to demonstrate as early as 1957 what the maximum sustainable yield of yellowfin tuna in the eastern tropical Pacific should be. Overfishing could now be easily recognized. When overfishing occurred it was recognized and an international catch quota was first applied on September 15, 1966. Quotas have been applied since, and as a result the stocks of yellowfin restored to optimum level. This not only bore witness to the soundness of Schaefer's studies but it demonstrated that a high seas pelagic fishery could be successfully "managed." This was no mean achievement.

In 1962 Schaefer was appointed professor of oceanography and director of the Institute of Marine Resources of the Scripps Institution of Oceanography. From this post he was "borrowed" for two years (1967–69) to serve as science adviser to the Secretary of the Interior. It was during his tour of duty as science adviser that he underwent open heart surgery to repair the damage to that organ which had started during the war. He never really completely recovered from this operation.

Of Schaefer's innumerable extracurricular professional activities only a few can be recorded here to bear witness to the great confidence that was placed in the scientific capabilities and integrity of this extraordinary man.

From 1956 to 1960 he was a member of the National Academy of Sciences–National Research Council's Committee on Effects of Atomic Radiation in Oceanography and Fisheries, and a member of the NAS-NRC Committee on Oceanography 1957–67 (chairman 1964–67).

He served as Expert on Fisheries on the Secretariat of the United Nations–International Conference on the Conservation of the Living Resources of the Sea, Rome, 1955; and on the U.N. Secretariat of the International Conference on the Law of the Sea, in Geneva, 1958.

He was the NAS member of the Advisory Board of the National Oceanographic Data Center 1960–63; member of the Latin American Science Board; chairman of the Standing Committee on Marine Sciences of the Pacific Science Association 1962–66; member of the Expert Panel for the Facilitation of Tuna Research for FAO of the United Nations; member and chairman (1965–66) of the (California) Governor's Advisory Commission on Ocean Resources; member of the Advisory Committee on Fisheries Oceanography, Department of State; consultant, President's Council on Marine Resources and Engineering Development 1967–69; member, California Commission on Marine and Coastal Resources; etc.

Although he was never really well after his serious heart operation in

1968, he did not spare himself. He underwent a second operation in 1970, but continued to fail in health. He took the death of his lifelong friend Wib, which occurred on June 25, very much to heart, but he continued his grueling pace. He insisted on attending a meeting of one of his numerous committees in Japan in early July, and soon after his return in a very weak condition on July 26, he followed his friend.

Chapman was sixty, Schaefer fifty-eight. Although both men knew they were very ill, they were defiant and unafraid. Their plans already formulated took them far into the future and at an ever increasing pace as if they knew that their time was short. Dylan Thomas, the fighting Welsh poet, recognized their breed.

> Do not go gentle into that good night,
> Old age should burn and rave at close of day;
> Rage, rage against the dying of the light.

The brave men died at the peak of their multitudinous activities. It is the way that they would have wanted to go.

J. L. KASK
San Diego, California

John L. Kask is retired. His professional career has included employment with the Biological Board of Canada, Pacific Salmon Commission, Halibut Commission, and California Academy of Sciences. He served as chief biologist, Food and Agriculture Organization Fisheries Division; assistant director, United States Fish and Wildlife Service; chairman of the Fisheries Research Board of Canada; and director of the Inter-American Tropical Tuna Commission. During World War II he was assistant chief, Fisheries Division, and chief of the Fish Resources and Research Branch of the United States Army GHQ, SCAP, Japan. He has served as commissioner with the International Commission for North Atlantic Fisheries, the Inter-American Tropical Tuna Commission, and the International Whaling Commission. He has acted as Consultant to the United States Department of State and the governments of Costa Rica and Iran, and has served on innumerable advisory committees and councils. He was a lifelong friend of W. M. Chapman and M. B. Schaefer.

Foreword

The papers published here were originally delivered at seminars held during the 1970–71 academic year at the University of Washington. This volume, like the several which preceded it and those which are planned to follow it, is a product of a continuing discussion among concerned students and scholars of resources and environmental issues, now in its tenth year at the university. The organizing theme of this effort, sponsored by the university's Graduate School of Public Affairs, has been to bring together persons with diverse interests and professional training to look at critical public policy issues in the development and management of natural resources and in protection of environmental quality. From the beginning, our objective has been to make the books reflect the spirit of the seminar itself: true interchange around a central theme by serious students of different problems.

It is entirely appropriate that a public policy seminar series originated in 1962–63 with fisheries as its central theme should produce a book ten years later which might be entitled "The Fisheries Revisited." It is also appropriate that this volume should be dedicated to Wib Chapman and Benny Schaefer, for the men who conducted these seminars in their name represented with them a "company of colleagues" of the type which underlies all interdisciplinary studies.

The substantive theme recurring in these papers is the need for policies and for institutions which can act wisely in the exercise of man's trusteeship of the riches of the sea. What emerges is a picture of a pluralistic decision-making system with many weak elements and some strong ones, with some interests overrepresented and some not represented at all, and of a growing body of scientific and technical knowledge that has clearly outrun both our understanding of the decision-making process and the capacity of our institutions to make full use of our opportunities for better resource management. The careful study of complex technical problems, their public policy ramifications, and, above all, the political and social mechanisms through which they are resolved is the essence of our concern with natural resource usage and environmental protection.

The papers also reveal a trend to be found in almost every field of human enterprise—the internationalizing of great issues and concerns The search for international political institutions that can express the needs of the whole globe and the creatures on it is just as desperate and poignant in the context of tuna and salmon fisheries or in ocean pollution as in the forum of nuclear weaponry and national power.

BREWSTER C. DENNY
Dean, Graduate School of
Public Affairs
University of Washington

Contents

Introduction

Recent technological acceleration and the diffusion of wealth and power among nations have stimulated substantive international reconsiderations of the allocation of ocean resources. These reconsiderations are accompanied by a variety of decisions which are, or will be, cast into national policies. At the same time, the dynamics of the international system must of necessity generate a re-examination of national policies regarding the oceans. Many of these international and national ocean policies involve, to a considerable extent, the fisheries.

The forecasting of the nature of future fishery policies is of considerable interest. The interest derives not only from any intrinsic values that the fisheries may have, but also from the likelihood that the fisheries will provide the substance for significant debates at the next Law of the Sea Conference. The shape of the fishery-related policy will be influenced by historical precedents and the attitudes toward them. In order to forecast future fishery policy, it is necessary to appreciate the history of the past as well as the attitudes of the present and the techniques of the future. We have accordingly provided a series of papers which somewhat eclectically record views on the history of fishery policy as well as some of the thoughts, the concepts, and the attitudes that will contribute toward shaping future policy.

The series of papers begins with Roy I. Jackson's discussion, offered from the vantage point of an internationalist. He sets the fishery problem in the context of the world food problem in an increasing population; an environment which is being inexorably modified; and an ever-increasing need to manage the stocks of fish effectively. Effective stock management and constructive approaches to slow accelerating environmental modification are constrained by conflicts in viewpoints among the states and among various interest groups. A concise history of fishery conflict and an analysis of the causes of conflict and some of the solutions that have been applied to at least ameliorate disagreement is given by Hiroshi Kasahara. The seriousness of the need to resolve these conflict areas is demonstrated by Donald L. McKernan. He points out that the growth in world fish catch shows signs of declining and yet management of the stocks in a broad sense is frequently of limited utility because the "management" is generated by organizations that were created to solve problems prevalent decades ago and are not responsive to contemporaneous fishery problems. Ambassador McKernan outlines principles required for resolution of conflict areas. The development of these principles rests in the social sciences as well as in the subtleties of population dynamics and other sciences more traditionally related to the fisheries.

In the area of social sciences, William T. Burke outlines some of the prolegomena for a new legal order in the world oceans in terms of the objectives involved in maintaining the *status quo* or deriving a change in the order. He analyzes in some detail the various components that will contribute to a modification of the law of the sea. An important component toward any modification is the criteria by which we judge alternative ways of achieving change. Foremost among these criteria are those that have an economic rationale, which James A. Crutchfield discusses in another social science related paper. He describes the drift from the maximum sustained yield criteria as part of a general discussion of the application of economic theory to fishery problems.

There are several actual fisheries where all of these problems are dramatically displayed. One of these is the cosmopolitan fishery for tuna. James Joseph reviews the "world" tuna fishery and the institutions that have been developed for the purpose of monitoring or managing the stocks of tuna. Joseph reflects upon the inadequacy of regional commissions and calls for a single world commission. Not only will we have to consider undertaking the management of certain groups of species, such as tunas, from a global point of view, we will need to undertake the management of entire large regions of the ocean, not only with respect to all species of fish, but also with respect to the interaction among the fisheries or potential fisheries and the social and economic needs of the coastal inhabitants. Such a large multifaceted undertaking is, in fact, beginning. It is the Indian Ocean Program of the Indian Ocean Fishery Commission. The rationale of fishery development in the Indian Ocean and the program of activity are described by John C. Marr.

The dynamic flux of views in the international community concerning the utilization of ocean resources elicits a variety of requirements among the various states. An effective regime for the conservation of the world's fishery resources will have to be built upon an appreciation of these requirements, which will be difficult to gain because of differing attitudes among the states. Examples of the bases for these varied requirements can be derived by comparing several papers in this volume. In particular we have J. L. McHugh's paper which describes the relation of United States fishery policy development to Jeffersonian democracy and shows some of the management patterns that have evolved in a developed country. The interactions of the attitudes toward fishery management of the developed countries with those of the so-called less developed countries can be seen in J. A. Gulland's study of the contrasting viewpoints held by the various nations. The rationale for concern for fisheries in the development of national policy particularly in some developed countries may be difficult to grasp. The difficulty is obviated by William Terry who makes a pressing case for applying care in fishery policy development because it is this policy which will preserve future resource utilization options.

The discussions clearly show that we are at a point in time when it is highly appropriate to take stock of where we are with respect to fishery management, to evaluate some of the frequently glossed-over faults of the past, and to ask questions that challenge society to an improved management of our fishery resources. This is the spirit of Peter A. Larkins' paper. He offers the hope that fisheries science will lead the way in evolving new methods of management. It is clear that these must involve much broader perspectives than have heretofore been appreciated. This broadened view requires an increased utilization of modern informational methods, and clear philosophical foundations. With respect to a broadened view, William F. Royce touches on some of the social and political aspects of the fishery problem that are not treated by the other authors. One of the modern techniques in information science is systems analysis. Brian J. Rothschild discusses the application of systems analysis to the problems of fishery policy development. The need for a systems approach in resource management is reflected by the endless scientific discussions that accompany fishery management problems. But science is a truth seeking process whereas management is a decision process. This is discussed in the paper by Dayton L. Alverson. Systems analysis requires the use of a variety of tools, many of them in the area of operations research. Effective utilization of these tools requires computer technology, aspects which are discussed by G. J. Paulik. Finally, all of our work must be based upon an underlying philosophy. G. L. Kesteven outlines such a philosophy in a paper that relate the questions of management to the necessary revolutions in biology and in ethic.

Thus, we have brought forth a collection of papers on fisheries problems which reflect the breadth and complexity of these problems.

The challenge lies not in simply describing this complexity, but in developing explicit policy which will contribute to the resolution of the complexity, thus enabling more effective management of our fishery resources. It is sad that we will move more slowly toward this goal without Wilbert M. Chapman and Milner B. Schaefer, as well as Vernon Brock, Hiroshi Nakamura, Clarence Pautzke, Elton Sette and Gerald J. Paulik.

B. J. R.

World Fisheries Policy

Multidisciplinary Views

1

Fisheries and the Future World Food Supply

ROY I. JACKSON

In the century just past we humans have changed our numbers, our way of living, and our environment faster than in any comparable period of history. The rate of change continues to accelerate. Our fisheries problems are sometimes regarded as minor among growing world concerns. But they are ecological concerns, and are part of what Garrett Hardin calls "the tragedy of the commons." The commons is the earth, and the essence of the tragedy resides in the remorseless consequences of not obeying nature: the ever-expanding world population, accumulation of its wastes, and disruptions from expanding technology.

Nature and man's fate are inseparable, and future fishing cannot be evaluated apart from the world of men or from the environment. We must estimate how many people may exist in the future. It is imperative that we recognize that both the quality of human existence and the quality of the environment in which fish live will be determined largely by what we do to the total environment. We must determine how much food, particularly animal protein, the future population will require, and consider possible new foods. We can make informed guesses about the sizes and kinds of future catches and which sea and freshwater areas will produce them. None of us believes any longer, if any of us ever did, that aquatic resources are limitless.

The total picture is obscure and many believe that it is short on hope, but we can and must develop some lines of action for the future.

Roy I. Jackson is deputy director-general of the Food and Agriculture Organization of the United Nations. He was previously director of the Fisheries Division of the FAO, and on the staffs of the International Pacific Halibut Commission and the International Pacific Salmon Fisheries Commission. He also served as the executive director of the International North Pacific Fisheries Commission.

This paper was adapted from the keynote address presented at the American Fisheries Society Centennial Celebration, New York City, September 14, 1970, and subsequently published in *Transactions of the American Fisheries Society,* 100 (1971): 151–58.

The Future Population

Five thousand years ago, when our written history began, the world contained fewer people than New York does today. One hundred years ago, the world population reached 1.4 billion. Today we are 3.7 billion. Many of us have lived through one doubling of the world's population and many of us will live through a second—by the year 2007.* In another hundred years, if the 1968 rates of increase continue, the world will have 29 billion inhabitants. The youngest of us may have surviving children at that time and many of us will be represented by grandchildren.

There are enormous differences in population distribution by countries and by regions. Today 28 percent live in the so-called "developed regions," which include Europe, the Soviet Union, North America, Oceania, and Japan; and 72 percent live in the so-called "developing regions," which include the Far East (excluding Japan), the Near East, Africa, and Latin America.

Undoubtedly these differences will be aggravated. Some countries will be less interested or less successful in limiting their populations. If present regional trends continue for the next hundred years, 10 percent will inhabit the developed countries and 90 percent the developing regions.

The demographers' predictions generally stop at the year 2030, because unknown changes in the rate of increase are expected. If desirable progress in health and social justice continues, the population could, for a time, become even larger than present trends indicate. But rates of population growth *will* eventually decrease. Wars, disease, and famine will reduce survival in proportion to our failure to limit births. This aspect of the tragedy of the commons has no purely technical solution.

The Future Environment

Our multiplying population and advancing technology combine to make us the most influential part of the earth's ecosystem. We are responsible for the most precipitous changes, both damaging and beneficial. Peoples of advancing cultures have always prided themselves in being "conquerors of nature." We have been paramount among them, and we are increasingly prodigious consumers as well.

The word "consumers" is used advisedly, because it is as consumers of natural resources, renewable and nonrenewable, that we must see ourselves. On the average, with massive inequalities, we enjoy higher stan-

* Based on the United Nations medium assumption for population growth rates.

dards of nutrition, health, and shelter than society has ever known before. To acquire these essentials—and many nonessential amenities—for our greatly increased numbers we are greatly increasing our per capita consumption of the resources of the earth that are essential for our life processes.

We know that every stock, every living population, if its numbers are to be sustained, must come into balance with its environment and its food supplies. As we now live, our renewable living resources depend on nonrenewable natural resources: oil and coal, and other minerals, environments, and ecosystems. Since they include self-reproducing organisms, ecosystems are not customarily listed among nonrenewable resources. But once destroyed, their former structure cannot be reconstituted. The supply of all these nonrenewable resources, including conventional energy sources, is limited. All the food that is produced or caught or distributed by modern methods costs a great deal in energy, whether this be used to make steel, or to fuel chemical fertilizer plants, tractors, or fishing boats. Therefore, we must be aware of the eventual consequences that can come from exercising our clear capability of looting the commons.

We create wastes in proportion to our use of resources. Wastes that are not neutralized, stored, or put to beneficial use become pollutants. The quantities and varieties of pollutants increase exponentially in relation to our population growth. In the United States the rate of increase of industrial wastes is three times as high as the rate of population increase, and there is three times as much industrial pollution as domestic pollution.

The waters where our inland fisheries and the artificial culture of fishes take place are highly sensitive to pollutants. Our coastal fisheries are in zones that are first to be affected by the outpourings from the land. The Federal Water Quality Administration lists five major kinds of pollution in coastal areas: bacterial contamination; decomposable organic materials that deplete dissolved oxygen; pesticides, herbicides, and toxic wastes from chemical manufacturing; materials that act as fertilizers for some life forms at the expense of others; and inert materials that fill invaluable estuarine areas and smother benthic life forms.

To this list we can add thermal pollution, oil spillages, and, for heightened drama, dumpings of leftover mustard gas in the Baltic and nerve gas in the Atlantic. Even what we cast into the air finds its way into the sea.

We know that the blight is broad and spreading. The historic Rhine has been a historic sewer for a long time. The deeper layers of the Baltic have much less oxygen and much more phosphorus than at the beginning of the century. The Soviet Union, in spite of its vast area and centralized authority, finds that its rivers, lakes, and coastal waters are rapidly deteriorating. More examples, of varying scope, can be found in every continent and on most inhabited islands of the world.

In his keynote address to the Ninety-third Annual Meeting of the American Fisheries Society, Justin W. Leonard spoke of the "ecological illiterate," those who plan and operate our technological society but think that food comes from the supermarket and water from the tap. "Ecology and natural history," he said, "have become old-hat. They aren't quite respectable anymore." Times have changed again. Everywhere we hear loud alarms, read passionate convictions, and see action to protect or restore the quality of our environment.

There is still a danger, however, that the subject can become old-hat again if we let up on either investigating or publicizing the continuing issue. We must continue to examine and report the more obvious effects of pollution. But as specialists we should continue to uncover and relate the less obvious effects of conflicting uses of water as well.

The effects of altered water flow regimes on snails can be taken as examples. In the Potomac River the oyster drill, a snail, is killed each year during spring freshets. This permits an extension of oyster culture up the estuary. Proposed dams on the Potomac will regulate spring flows, the drills will not die, and fewer oysters will live to be eaten by man. In the Nile River region perennial irrigation replaces seasonal flooding. This also favors increases in a snail, the intermediate host of a parasitic worm. The worm causes extremely debilitating schistosomiasis disease in man. In upper Egypt the new Aswan High Dam development may increase the intensity of infestation. At the same time it may seriously reduce the eastern Mediterranean sardine fishery while providing an impoundment for freshwater fish.

In the case of the Nile, man faces the immediate problem of deciding whether the increased electrical power and starch and freshwater protein are worth the increased disease and decreased marine protein. A less immediate but perhaps more important problem is whether the impoundment behind the dam, which has a lifetime probably measurable in decades, has a long-term value equivalent to the marine environment that will be affected.

An economist might calculate, by using widely accepted economic value criteria, that the most profitable use of the Nile can be obtained by damming it, and that the most profitable use of the Rhine is as a sewer. Also some apparently calculate that the most profitable way to obtain oil from Saudi Arabia is first to flame off the natural gas and add a bit more carbon dioxide to the already overburdened atmosphere. I disagree with most conventional profit valuations. They include private costs but overlook social costs, and lead to uses that would deny the earth to future generations.

Besides the continued expansion of present multiple uses of water, we can expect more kinds of uses, especially in the ocean. The ocean floor is crisscrossed with cables, and pipelines are following. Oil derricks now line the horizon in many nearshore areas, and underwater oil storage tanks may become common. Mining from the sea floor will certainly increase. Except for the fact that boats must dodge and gear may foul, these physical structures and activities do not conflict greatly with fishing. Accidental release of oil from wells, huge tanks, or pipelines, however, could cause much pollution. Other uses of the sea might, for example, require diversion of currents to change weather patterns. This kind of activity should be approached very cautiously. It could, in some ways, be very useful to terrestrial man while very harmful to aquatic systems.

The Future Fisheries

The fisheries can do much to help meet the continuously increasing demands for food. Estimates indicate that the world catch of fish today could supply about 70 percent of the *animal* protein requirement of the present population. This figure is subject to many qualifications, and it should not be interpreted to mean that fish *does* supply that much of human needs. More than half of it is consumed by livestock, and the world distribution is very uneven. But it shows how important fish protein could be in the world diet.

The record of the quarter century from 1945 to 1970 is encouraging for the decades immediately ahead. Since 1946, catches have increased about 6 percent per year—considerably faster than the world population—to reach a total of 64 million metric tons in 1968. Of this total, about 7 million tons came from fresh water. Forty-one percent of the marine fish catch is taken in the Atlantic, 55 percent in the Pacific, and only 4 percent in the Indian Ocean. Divided another way, 54 percent is taken in north temperate waters, 29 percent in south temperate waters, and 17 percent in tropical waters. The north-south division shows the great expansion of fishing beyond the northern waters, where 73 percent of the total was taken in 1958.

Many estimates have been made in the last twenty-five years about how far this expansion can continue. These range from extremely conservative estimates, which have been reached and surpassed by the actual catches within a few years after being made, to as high as two billion tons—some thirty times the present.

A detailed estimate of the world aquatic potential was completed by the Food and Agriculture Organization in 1970 as part of the Indicative World Plan for Agricultural Development. The FAO study produced several figures, because much depends on what is included in the potential. The largest possible harvest sources are the plants of the sea and freshwater fisheries. Ocean plant production is fairly generally agreed to be in the range of 150–200 billion tons per year. Man's annual harvest could approach this production if it were technically feasible to catch and process the very small plants and animals at an economic cost. Although the technological and economic possibilities for the year 2070 are not predictable, no method for economically harvesting or using a significant proportion of this material is even conceivable at present.

The FAO study succeeded in making estimates for nearly all those animals that now support major fisheries: whales, large pelagic fishes (tunas, billfishes), medium to large demersal fishes (cods, flounders, seabreams, etc.), and shoaling-pelagic fishes. Under ideal conditions of exploitation, these together could provide catches of about 100 million tons. But the limit of the "traditional" ocean fish (excluding squid and other molluscs) is likely to be reached in the 1970s. Even this may be optimistic, because it would require that we obtain the maximum catch from all stocks. Preliminary figures suggest that the 1969 world fish catch was somewhat less than that of 1968—the first decline since FAO started collecting comprehensive world statistics nearly a quarter of a century ago.

The familiar types of crustaceans (shrimps, rock lobsters) could provide somewhat over 2 million tons per year. Large quantities (1.2 million tons in 1968) of squid, cuttlefish, and octopus are being caught. No estimates could be made of their potential, but since various species of squid are found commonly in all parts of the ocean, their potential must be large.

No projected estimate was made for the other molluscs (clams, oysters, mussels) of which the 1968 world harvest was 2.2 million tons, because the possibilities for increased harvest come more from cultivation than from natural production.

Finally, estimates of potential catches from the sea must include the smaller but exceedingly abundant animals, such as the krill (euphausids) of the Antarctic and the lantern fish (myctophids). At present it is not considered possible to harvest these economically, though the Soviet Union seems close to using krill on a commercial scale. The potential of these small animals is vast, probably several times the present world catch of all fishes.

The yield of freshwater fisheries for 1968 was 7.4 million tons, about 11.5 percent of the total world production. This excludes the very large subsistence and sport fisheries, estimated to be at least half as large as the recorded commercial catch. Inland fisheries could provide much more food than they do now. Their future depends largely on the prevention of further deterioration in water quality, and on the improvement of those waters that are already despoiled.

The catch from large inland lakes and rivers probably could be doubled, but the major increase from lakes and reservoirs will come from the smaller bodies of water where management techniques can be applied. In these waters up to five times the present catches seem possible. Controlled culture, however, is our greatest opportunity for increasing fish production. The larger lakes produce about 5 kg/ha; the smaller lakes produce up to 150 kg/ha. Managed ponds in tropical and subtropical areas commonly produce 1,500 to 2,000 kg/ha, and under very intensive management 6,000 to 7,000 kg/ha.

Generally fish are fed supplemental materials that are not now consumed by humans, and some convert vegetable proteins into animal protein, including all ten essential amino acids, very efficiently. For example, some work in the United States has shown that channel catfish have a feed conversion of 1.3 (that is, it takes only 1.3 pounds of feed to produce 1 pound of flesh). By contrast, beef cattle have a feed conversion of about 16.

The Lines of Action

Our action must be fundamental. To deal with effects without also dealing with causes is inadequate and superficial. What M. King Hubbert has written applies to fishermen as to all men. As he sees human history, the period of rapid population and industrial growth that has prevailed during the last few centuries is an abnormal, brief, transitional episode. He foresees a period of nongrowth that will pose no insuperable physical or biological problems but that will entail a fundamental revision of our current economic and social thinking.

Future nongrowth of the human population is a certainty. When this will occur, at what maximum number, and through what mechanisms —barring natural catastrophe—depends entirely on us humans. This is not just a problem for Asians, Africans, or Latin Americans. It must be faced by every one of us in his own neighborhood.

In many of our activities, we, the technologically developed cultures in particular, follow the archaic approach to the problem of the commons, that of free and unlimited access. If this approach is justifiable at all, it is justifiable only under conditions of low population density. As the human population has increased, the commons has had to be abandoned in one aspect after another. Traditionally we have treated the air we breathe and the waters of the earth, along with their inhabitants, as commons. This is changing, and it must change more radically, and soon.

Properly oriented changes can only occur where there are perceptive and knowledgeable persons to show the way, to monitor, and to be watchdogs. Most natural history movements have addressed themselves to terrestrial communities; there has been a dearth of guardians of the waters. Meanwhile fishery biologists have to a large extent allied themselves with conventional exploitative processes. Even personally we have not been sufficiently appalled by the demise of environments and the extinction of aquatic species and communities.

Within the framework of the problems we must face as citizens of the world and as general watchers of the waters there are particular fisheries problems that we must face as fisheries scientists and administrators. The pressing problems of the world fisheries, at least for the early part of the coming century, are three: (1) to manage the limited resources of "traditional" fisheries in the most effective way; (2) to develop fisheries on the large resources of less familiar species of fish; and (3) to increase cultivation, especially of species (for example, some molluscs and freshwater fish) that feed directly on plants.

We must be particularly concerned with proper management. The past record of management shows the effects of our reluctance to abandon the commons with respect to fisheries. On the high seas, the Antarctic whales were rescued on the limit of commercial extinction—possibly absolute extinction of blue whales. Where only one country is concerned the record often is not much better—the California sardine is an example. There have been successes, however: effective conservation of the whales in the Antarctic is beginning, and fishing is controlled in several major fisheries.

Three forms of jurisdiction have been proposed for future management: wide extensions of fishing limits, to place most fish stocks under national jurisdiction; direct international, United Nations, control of high seas resources; and expansion of the present pattern of regional international fishery bodies and commissions.

Management must start with control of the individual fisherman. We generally assume that anyone should be free to fish on the high seas, or any national within his own territorial waters, so long as he does not use obviously damaging methods like poisons, explosives, or devices that catch immature fish. Fishing is constrained by restricting the effectiveness of each fisherman—by explicitly prohibiting the most effective gear, or by closing areas or seasons.

The traditional assumption, that fishing is free to all, is unrealistic and has led to inefficient resource management. Each fish stock is limited. If it is accepted that fishing is a privilege, not a right, then one likely method of controlling excess fishing is to charge for this privilege. This control could be accomplished if the payment or license fee were in proportion to the privilege conferred. In several major fisheries, for example, Pacific salmon, the gross value of the catch greatly exceeds the basic cost of harvest, sometimes several fold. There the right to fish with the most efficient gear might be worth up to 80 percent of the gross value of the catch.

Who gets this license fee is a matter of jurisdiction. Inshore it could clearly go to the coastal state; offshore it might, under one scheme, go to the proposed international agency.

If we optimistically assume that improved management practices are instituted, it is possible to visualize the ocean fisheries of the early twenty-first century. On the fishing grounds that are familiar today—the Grand Banks of Newfoundland, the anchovy fishery off Peru—the fishing vessels will be fewer. They will be helped by a flow of information on the distribution of fish, and on weather and water conditions from satellites and buoys. These vessels will make large catches and pay substantial license fees. Some of these fees will be used to provide the satellites and other information systems, and the scientific research on which the management is based.

The other major sea fisheries, dominant in weight but probably not in monetary value, will be in the Antarctic on krill, and along the major upwelling systems on the small lantern fish and other animals. If the traditional stocks were harvested efficiently, it would become possible for men and vessels to be diverted to these less familiar stocks as well as to stocks like squid and whiting that are not being used to the extent they might be.

The development of the technology to harvest and use the less familiar fish will demand initially the resources of the richer developed countries, and will at first be of less concern to the developing countries, for whom FAO has a special responsibility. But there are already shifts in emphasis and interest in fisheries from the highly developed countries to the intermediate nations. Because labor costs are less and other economic opportunities are fewer, the new fisheries are likely to be developed and used by developing countries in the long run. This is probably economically desirable for the world as a whole.

Fish flesh contributes about 11 percent of the animal protein now consumed by man. This percentage should increase considerably in the future. In order to do this we must develop more efficient catching methods and provide adequate transport and processing, especially in the developing countries. For example, the control of insect infestation of fish products in Africa could double the amount of fish reaching consumers. Above all we must alter our eating habits. Thousands of tons of good protein are unused because even people who do not already have adequate protein in their diets refuse to eat all but a few traditional species.

Controlled culture of marine and freshwater species is a great opportunity for increasing production. Through such techniques as the raft methods developed for mussels in Spain, the possibilities for shell fish seem very large. Running freshwater cultures are highly efficient in converting fish food to human food. We must develop and apply these techniques to commercial production. Controlled culture is limited by economic consideration rather than natural productivity, at least until we run short of nonrenewable resources.

I have stated earlier that the total picture of world problems is obscure and may be short on hope. But there are some developments from which we can take heart. The population problem is at least recognized as the central theme of the tragedy of the commons. And in highly developed countries the environment has the limelight. There is action as well as talk.

National and international actions are having some positive effects on fisheries management, and initiatives that affect fisheries resources have multiplied in recent years. The government of the United States has established a National Oceanic and Atmospheric Administration to deal with environmental problems, including fisheries research and administration. *Pacem in Maribus,* the convocation on the oceans held at Malta in June 1970, highlighted the growing competence of the world to exploit the oceans' resources. The need to strengthen present measures and introduce new ones to preserve renewable resources and make beneficial and equitable use of the others was clearly expressed. We in FAO are striving to strengthen the growing network of regional international fishery bodies. A third Law of the Sea Conference is expected to be held in 1973. Many other important initiatives could be added to this short list. A great value of the exploited ocean may be that it will encourage nations to regard one another as partners in world progress.

The Second World Food Congress, held in June 1970 in The Hague under FAO auspices, concluded that world food supplies can be adequate in the decades immediately ahead; so it appears that famine is not imminent. After all, animal protein, which includes fish, is an essential part of this food supply. It is both humane and very pragmatic that we increase production and that it ends up—preferably is produced—where it is needed most. This will require some changes in traditional economic practices, and it will certainly require that we eschew warfare, which brings famine and disease faster than any other human activity.

Whatever its portents for the far future, carefully applied technology is needed to grow or catch the foods that the growing population will require. In addition it could provide some new sources. Scientists apparently have isolated bacteria that require only methane to multiply. The bacteria are 50 percent protein. In Britain a plant to produce sixty thousand tons of protein per year, and a hydrocarbon-based yeast factory to produce four thousand tons per year, are being built. Several scientists have stated that before 1980 it will be possible to produce protein artificially from petroleum in unlimited quantities. These developments could become important in providing protein, so long as the limited supply of fossil fuel lasts.

This is my view of the near future. But to conserve nonrenewable resources and preserve the environment for ourselves and the renewable resources in the increasingly crowded and technically complicated world of the next hundred years will be a formidable as well as exciting job. Fisheries workers must be responsible for the aquatic phases of that development. We must remain aware that even a century allows us too little foresight. The long-term life-carrying capacity of the planet must be our most vital concern. There is still time to act on this concern, and fortunately nature eventually corrects many of our mistakes—if we behave rationally.

September 1970

2

International Fishery Disputes

HIROSHI KASAHARA

International problems of fisheries are not new. Major disputes over fishery matters date from the early seventeenth century when England made a unilateral claim to restrict herring fishing by Dutch fishermen along the coasts of Scotland and England. The issue was of great economic importance: the gross value of the Dutch catch amounted to 2 million sterling pounds, while the total English commodity export was 2.5 million sterling pounds (Leonard, 1944). The English made various attempts to exercise jurisdiction over Dutch fishermen with a mixed result. The Dutch considered this a violation of their rights established over a long period of time. The dispute was not settled by negotiations but was solved as a result of repeated wars on the Continent, from the mid-seventeenth century to the early eighteenth century, which exhausted the Dutch. The British fisheries began to dominate in Europe. The concept of the territorial sea had not been established, although the Dutch appeared to recognize that the English had some special rights to fisheries in waters within some distance from the shore. The question of conservation played no part.

The North Atlantic Coast fishery arbitration was one of the earliest international problems dealt with by the United States after its independence. Rights to conduct fishing and fish processing activities in Newfoundland were granted to American fisherman when the British in the colony were not aware of the rich resources along their own coasts. As the potential of these resources became known, the British government attempted to institute regulations discriminating against the American fishermen. In addition, the Newfoundland authority thought that the control of the fisheries could be used as a means to relax trade restrictions on products exported to the United States (Leonard, 1944).

Hiroshi Kasahara is director of Program Coordination for the Department of Fisheries, Food and Agriculture Organization. He was formerly professor of fisheries at the College of Fisheries, University of Washington, a senior consultant of the United Nations Development Program, a research biologist with the Nippon Suisan Company Research Laboratory, a member of the Central Fisheries Advisory Council of Japan, and was employed by the Research Division of the Japan Fisheries Agency.

The dispute appeared to be resolved by an agreement of 1818 which redefined the rights of the American fisherman. But controversies occurred again and were dealt with by temporary agreements for a number of years. The issue was finally submitted in 1908 to an international arbitration tribunal, which recommended certain procedures to settle problems, including the formation of a mixed commission. Throughout these controversies, the question of conservation was not an important consideration. The main issue was of an economic nature arising from the recognition by Newfoundland, and later Nova Scotia, of the potential of the resources along their coasts (Leonard, 1944).

Another outstanding example of a dispute over inshore fishing rights was the Russo-Japanese controversy. Japanese fishermen started fishing, mainly for salmon, on Russian territory in the latter half of the nineteenth century. After the Russo-Japanese War, the two governments entered into negotiations on the question of Japanese fishing rights along the Russian coasts. The resulting treaty of 1907 (went into force in 1908) greatly facilitated expansion of Japanese fishing activities on Russian territory. The treaty recognized the rights of the Japanese to fish along the Russian coasts of the Japan Sea, the Okhotsk Sea, and the Bering Sea, including the entire coastline of Kamchatka, the most important salmon producing area in Asia. A large number of lots for salmon trap fishing were distributed, by auction, between the Russians and Japanese.

After the Russian revolution, a temporary arrangement was made until a new treaty was signed in 1928 after long negotiations. The new agreement modified the arrangements of the 1907 treaty, but the rights of the Japanese to fish along the Russian coasts were re-established. Many years of controversies and negotiations followed, as the Soviet government tried to expand the fishing activities of its nationals by reducing the proportion of fishing lots leased to the Japanese. The amount of Japanese fishing for salmon from the Russian territory was reduced drastically after 1941, when the Pacific war broke out, and the whole business ended in 1944.

While negotiations for the allocation of fishing lots were going on between the two governments, the Japanese began to engage in a mother-ship-type salmon fishery in waters off Kamchatka. More important was the development of drift-net and trap fishing based in the northern islands of the Kurile chain, which intercepted a large number of salmon on their way to the streams in the Kamchatka peninsula and other areas of Russia (Kasahara, 1961). It appears peculiar that such new developments, which no doubt had substantial effects on inshore salmon catches, did not result in a serious controversy during the prewar period. The question of conservation was not a matter of great concern, although inshore salmon fish-

---

He has held positions as assistant director of the International North Pacific Fisheries Commission and senior project officer for the United Nations Development Program.

ing was regulated by Russian law. Salmon fishing was of substantial economic importance to both nations, but national prestige and diplomatic considerations also played an important part.

Both the North Atlantic and the Russo-Japanese disputes concerned the rights of the nationals of a state to conduct fishing activities in the inshore waters of another state. Problems of this nature are now of relatively minor importance, except for those of historical rights resulting from the extension of national jurisdiction.

International problems arising from high seas fisheries, too, are not new. A major conflict developed in the early nineteenth century, after the Napoleonic wars, between English and French fishermen over the English Channel fishery, and resulted in a convention in 1839. The treaty recognized three miles as a general limit for exclusive national jurisdiction with an exception for the oyster fishery in the Bay of Granville. A joint commission was established to develop detailed regulations. This might be considered a forerunner of modern arrangements for international regulation of fisheries. Violations and enforcement problems made it necessary for the two governments to consider a new convention. Such a convention was concluded in 1869, but the French government failed to pass legislation to enforce the new regulations (Leonard, 1944).

An international convention for policing of the North Sea fisheries outside territorial waters was signed in 1882. It defined exclusive fishery limits and provided regulations mainly to avoid conflicts between trawlers and drift-netters. On the research side, mention should be made of the establishment, in 1902, of the International Council for the Exploration of the Sea (ICES) for the purpose of facilitating and coordinating fishery and oceanographic studies of the northeastern part of the Atlantic. It has survived the two world wars and is still going strong.

International disputes over the fur seal populations in the northern North Pacific started about twenty years after the purchase of Alaska, when the United States decided to enforce sealing regulations beyond its three-mile limit. North Pacific fur seals breed on certain islands, mainly the Pribilof Islands of the United States, the Commander Islands of Russia, and Robben Island of Russia (previously Japan), the Pribilof population being the largest. They conduct feeding migrations over large areas of the northern North Pacific and can be caught easily on the high seas. The American government took the position that all the waters of the Bering Sea east of the 1867 treaty line were those of Alaska territory, and actually seized vessels from British Columbia (British nationals in British Columbia had developed a large high seas sealing fleet to operate in the North Pacific and Bering Sea). The issue was finally referred to a tribunal of arbitration and an agreement was reached in 1893.

Meanwhile, consultations were also held between the Russian and

British governments, as well as between the Russian and the United States governments, concerning high seas sealing. By the early 1900s, it became obvious that the populations of fur seals had been so badly depleted that a broad international convention was required to preserve them. The United States took the initiative to convene a conference between the U.S., British, Russian, and Japanese governments in Washington, D.C., which resulted in a convention of 1911 for the preservation and protection of fur seals. Pelagic sealing was banned and a system of compensation for refraining from sealing was established. The convention was the first international agreement with main emphasis on conservation. It also established a unique system of revenue sharing. The seal populations, particularly the Pribilof herd, recovered very rapidly in the subsequent years.

The few examples mentioned above illustrate two things: international disputes over fishery matters have a rather long history; and, contrary to the widely accepted notion that most of the past fishery agreements have dealt only with the question of conservation, some of the early agreements quite squarely faced problems of a political or economical nature. Conservation has indeed been a basic theme for most of the international fishery agreements concluded in recent years. But other features, such as the allocation of resources or catch, or the special rights of coastal states, have also been just as important in most cases, whether or not they are explicitly written in the provisions of the conventions concerned.

Thus, the main feature of the North Pacific fishery treaty is "abstention," which is a form of resource allocation. The establishment of quotas for Japanese high seas fisheries is at least as important as conservation in the Northwest Pacific fishery treaty. The Japan-Korea fishery treaty, too, concerns itself with the allocation of resources and catch, as well as the distribution of fishing effort. The Fraser River salmon treaty, the main objective of which is the preservation of salmon populations in the Fraser system, also provides for a fifty-fifty catch division. The three crab agreements (Japan-Soviet, U.S.-Japan, U.S.-Soviet) deal intensively with catch quotas and, in the case of the Japan-Soviet agreement, limitations on fishing effort as well as the allocation of fishing grounds. The major whaling nations of the world have been negotiating for the allocation of catch outside of, but in close relation to, the whaling convention. The Inter-American Tropical Tuna Commission (IATTC) has begun to face the question of national quotas. Even the International Commission for the Northwest Atlantic Fisheries (ICNAF) has been giving more and more serious consideration to the feasibility of establishing national quotas for certain species. The main purpose of practically all bilateral, executive arrangements that have been made in recent years is to resolve issues of immediate political or economic importance.

The frequency of controversies and the complexity of the problems involved have increased in the postwar period as the pace of fishery development quickens on a nearly world-wide scale, with more and more resources becoming subject to intensive exploitation.

Causes of Conflict

A brief analysis of some of the basic factors contributing to fishery disputes might be helpful. Many of these are applicable to domestic as well as international problems.

The complexity of the technical factors involved in the regulation of fisheries, domestic or international, is often underestimated. Fishing is still largely an activity to harvest wild stocks of highly mobile animals. These animals cannot be fenced in limited areas or marked for ownership. While the number of species used is limited in Europe or North America, up to several hundred species are fished and marketed in some other parts of the world, particularly East and Southeast Asia. Even in the North Atlantic, there has been a substantial increase in the number of species commercially exploited in the last two decades. All of these species differ from each other in their life history, distribution, migration, and responses to environmental changes, including man's intervention. Virtually all of them go through planktonic stages in which they are members of communities consisting of a much greater variety of organisms. Compared with those available for land animals, means to study marine animals are limited and inaccurate. On the other hand, the rate of renewal of most marine animal resources, except mammals, is so rapid that little is gained by keeping them unexploited. Most of them occur in wide areas, and a great variety of usable species may be found in the same body of water. A number of species can be caught simultaneously by the same type of gear, typically trawl nets, while the same species may be fished by several different methods. Fishing in one area often affects the abundance of fish in other areas; fishing for one species may affect the stocks of others.

It is extremely difficult, under these circumstances, to find social justification for establishing sole ownership for a fishery resource and to make institutional arrangements for implementing such a system. Partly for this reason, fishery resources have been used traditionally as "common property," which can be exploited simultaneously by more than one individual or economic unit. In the domestic laws of a few nations (Japan, South Africa, etc.), rights of exploiting certain fishery resources are given to particular groups of fishermen, or companies, or sometimes even individuals, for the purpose of limiting entry to each fishery. Some of the complex problems resulting from nationwide application of such a regime are outlined by Kasahara (1964). But even there, the resources are utilized as common property, without ownership, by those who are authorized to do so.

Further complications arise from the heterogeneity of the fishing industry. Even in technologically advanced nations, there still exist a large number of fishermen who catch fish with methods not so different from those used hundreds of years ago, while large companies dispatch their fleets to distant water grounds. Internationally, problems are even more complex. Fishermen from nations at greatly different levels of development may operate in the same fishing area, often using similar techniques and equipment. For example, the per capita GNP (World Bank data for 1965) among the nations fishing in West African waters ranged from $60 to $3,240. The need for, and interest in, fishery development also differs from nation to nation. Some countries, such as Peru, Norway, Iceland, South Africa, and a number of others, make fish products mainly for export, while in many other nations, particularly those in East and Southeast Asia, fish is the main source of animal protein, often one half or more of the total animal protein intake being from seafood. To many others, fishery development does not appear a matter of high priority, either because they have sufficient animal protein supply from other sources, or because they can buy as much protein food as they need (Kasahara, 1970).

The relative importance of different species found in the same general area, too, varies greatly from nation to nation. A good example is found in the exploitation of groundfishes in the Pacific waters off the United States and Canada. While Russians and Japanese exploit abundant resources of such species as pollack, yellowfin sole and other flounders, ocean perch, and hake, American and Canadian fishermen do not fish some of these at all and exploit others to a very limited extent. What might happen to the halibut stocks off Canada and the United States as a result of development of the foreign trawl fisheries is a matter of great concern to the Canadian and American fishermen, but not to the Russians or Japanese except for political repercussions they get.

The organization of the fishing industry differs quite widely between nations. Practically all fisheries of the Soviet and other socialist nations are fully controlled by the government, with distant water operations carried out by state enterprises. This control is reflected in the systematic way in which their industries have developed in the postwar period. Some of the nations of free enterprise, such as Japan and South Africa, control the industry fairly well, though not as fully as the socialist nations. Many others, including the United States, have little control over the number or the type of fishing vessels used in a particular fishery or the distribution of vessels among different fishing areas.

Some nations are in a much stronger position than others to expand their fishing in international waters as long as the resources are used as common property. While the former are capable of going to any place in the world where abundant resources occur and of marketing the products either domestically or internationally, others have to stay in coastal waters for technological, economic, or institutional reasons. The latter nations naturally seek legal or political protection against the activities of the former in waters near their coasts. Even within a country, some sectors of the industry operate in waters off the coasts of other nations, while other sectors are affected by foreign fishing in their coastal waters. The American tuna fishery, for example, occupies waters off the coast from southern California to northern Chile, as well as off West Africa, and is considered a most aggressive fishery. At the same time, the Northwest and New England fisheries are crying for extension of national jurisdiction to restrict foreign fishing in waters off their coasts.

Diversity in the historical background of these nations also presents problems. Those who have developed their modern fisheries only recently, as have many of the less developed countries, wish to increase their shares of the catch from an area or a resource; those who have been fishing for many years in the same area or for the same resource hope to maintain their shares. A number of other factors contribute further to the complexity of international fishery problems, but a full discussion of these is beyond the scope of the present paper.

Present Legal Framework

Negotiations for fishery agreements have been carried out on the basis of two considerations: generally accepted principles of the law of the sea; and the need for specific arrangements for specific resources or areas between the nations concerned. Let us examine where we stand in respect of these two phases.

Because of the complex nature of the resources, as well as the way they are exploited, and the diversity of national interest, there still does not exist an established set of legal principles for international regulation of fisheries that are acceptable to most nations and workable in specific situations.

The four conventions resulting from the Law of the Sea Conference convened by the United Nations in 1958 form only a very general and ineffective framework for the regulation of fisheries. The Convention on the Territorial Sea and the Contiguous Zone (entered into force September 1964) failed to include an agreement on the most crucial issue, that is, the breadth of the territorial sea and so-called "exclusive fishery zone." A second conference was held in 1960, but no agreement was reached. It has been a general trend in the sixties and seventies for nations to expand, through unilateral claims, their territorial seas or exclusive fishery zones. The three-mile club is already a minority. The majority of the nations now claim twelve miles (either as territorial sea or combination of territorial sea and exclusive fishery zone). Even if a nation does not officially recognize the twelve-mile limit, it is now difficult for her nationals to fish freely within twelve miles of the coast of a state claiming it.

A substantial number of countries claim zones broader than twelve miles and up to two hundred miles, for example, Argentina, 200 miles; Brazil, 200 miles; Cambodia, shelf to 50 meters including superjacent waters; Chile, 200 miles; Costa Rica, specialized competence over living resources to 200 miles; Ecuador, 200 miles; El Salvador, 200 miles; Guinea, 130 miles; India, right to establish a 100-mile conservation zone; Republic of Korea, varying distances; Nicaragua, 200 miles; Panama, 200 miles; Peru, 200 miles; Uruguay, 200 miles; Philippines and Indonesia, all waters within their respective archipelagos.

As of October 1970, only thirty-nine states are parties to the territorial sea convention, while the membership of the United Nations has expanded greatly since 1958. Relatively few nations claim a zone broader than twelve miles and enforce such a claim, and are mainly in Latin America. But the trend of extension of national jurisdiction is irreversible and is likely to be accelerated each time the question is debated on a broader international basis. Furthermore, no complete agreement exists as to how the baseline (the line from which the breadth of the territorial sea and the exclusive fishery zone is to be measured) should be drawn, resulting in varying claims for the extent of internal waters (Harlow, 1968).

Many international fishery disputes have arisen from unilateral extension of national jurisdiction. Among the more serious ones are: the United Kingdom versus Iceland; the USSR versus Norway; the United States versus Chile, Peru, and Ecuador; Japan versus South Korea; Japan versus Indonesia; Japan versus Mauritania; the United States versus Mexico; the USSR versus Argentina; Japan versus Australia and New Zealand.

The Convention on the High Seas (entered into force September 1962) defines the freedoms of high seas, of which the freedom of fishing is one, and codifies laws of navigation. Forty-six states are parties to it as of October 1970.

The third convention, Convention of the Continental Shelf (entered into force June 1964), has a membership of forty-five states as of 1970. The main purpose of this treaty was to define the extent of sovereign rights to be exercised by coastal states for exploring and exploiting nonliving resources of the sea bed and subsoil thereof.* A small number of nations succeeded in including some of the living resources with a somewhat ambiguous definition. The membership of this convention includes most of the major fishing powers, with some notable exceptions, and it is difficult even for a nonsignatory state to ignore its provisions. The convention has had substantial effects on recent fishery negotiations, including crab agreements in the North Pacific. It is possible that the definition of shelf-living resources subject to sovereign rights might change as a result of the proposed next Law of the Sea Conference, as will be mentioned later.

The fourth convention, Convention on Fishing and Conservation of the Living Resources of the High Seas (entered into force March 1966), makes it mandatory for member nations to enter into negotiations, if proposed, with a view to reaching an international agreement for conservation purposes. The coastal states must not be excluded from such arrangements even if they do not use the resources concerned. A machinery for arbitration is also provided. Most of the major fishing nations, however, are not parties to this convention, and it has never been used as a means to settle fishery problems.

Finally, most nations have diplomatic reasons for avoiding or minimizing international disputes over fishery matters. In spite of the great publicity international fishery problems tend to attract, most governments, particularly those of technologically advanced nations, are well aware of the fact that these problems are usually not of great economic importance to the nations as a whole. The weight of such a diplomatic consideration, however, differs from nation to nation, as well as from case to case.

The above clearly indicates that the present law of the sea does not provide a sufficient basis for coping with the international fishery problems we continue to face, and the need for more specific arrangements is obvious. A wide variety of conventions for specific resources and/or areas have been negotiated between the nations directly concerned, and a number of international bodies (mainly fisheries commissions) established under such conventions. Numerous bilateral, executive agreements of a more temporary nature, and of limited scope, have also been signed in recent years. Although agreements of the latter category do not involve the operation of international bodies and have to be revised from time to time, they are becoming an increasingly important part of the international regulatory regime for fisheries. The United States, for example, is

* The main issue in this respect is the definition of the shelf area subject to the convention. The present definition is perhaps a good one from a practical point of view but leaves much room for legal debate.

a party to nine international fishery conventions (including one for the Great Lakes) and, I believe, nine bilateral agreements of the second category, all dealing with specific resources and/or areas.

The scope and nature of these arrangements vary depending upon the biological characteristics and status of exploitation of the resources concerned, the interests of the nations in the particular fisheries, and above all the political circumstances under which such arrangements are made. The results are usually not entirely satisfactory to any of the parties involved, but generally contribute toward minimizing international disputes and avoiding the disruption of major fishing activities. Many of them are also useful for maintaining yields from the resources concerned at levels higher than they would be without such agreements. As critics of the existing arrangements put it, "They are better than nothing."

As there is no single set of principles widely acceptable, the network of these arrangements can become very complicated. Figure 1 and Tables 1 and 2 show, in an oversimplified way, the legal framework in which the fisheries of nine different countries operate in the North Pacific. The actual situation is much more complicated; there are still no diplomatic relationships between some of these countries.

TABLE 1

Membership Status of Four Conventions for Nine Countries Bordering the North Pacific as of October 1970

Countries	Conventions			
	Territorial Sea & Contiguous Zone	High Seas	Continental Shelf	Fishing & Conservation
United States	+	+	+	+
Canada	−	−	+	−
Mexico	+	+	+	+
USSR	+	+	+	−
Japan	+	+	−	−
South Korea	−	−	−	−
North Korea	−	−	−	−
China (Taiwan)	−	−	+	−
China (Mainland)	−	−	−	−

There also exist a number of commissions, committees, and councils established by the Food and Agriculture Organization of the United Nations (FAO) to deal with problems of fisheries, but practically none of them have regulatory functions; they may be regarded as advisory bodies. But FAO has also been active in promoting the conclusion of regional fishery conventions and, on two occasions, has prepared a draft treaty and called a conference of plenipotentiaries to conclude a convention.

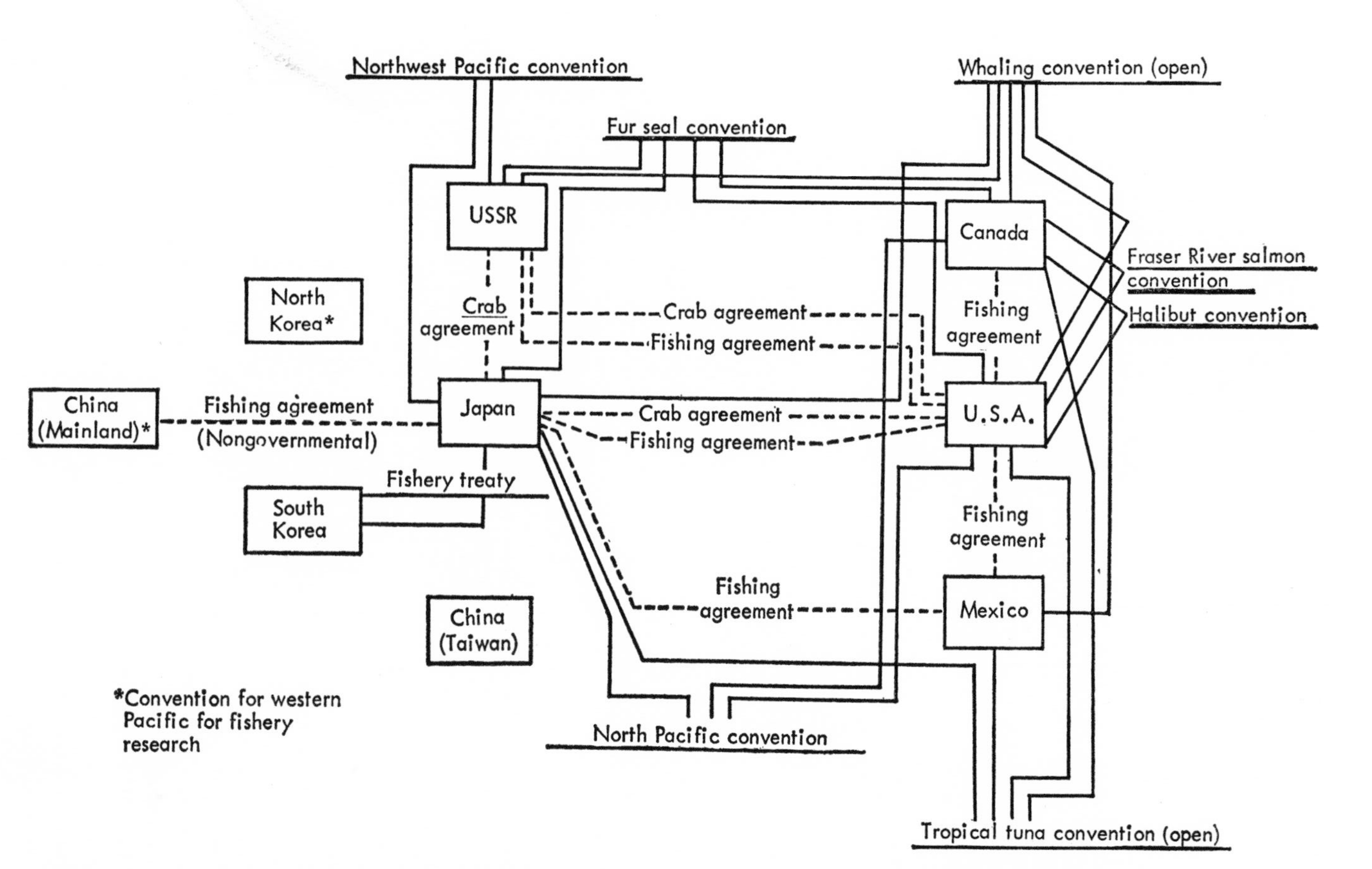

Figure 1. Fishery conventions and agreements in the North Pacific (as of October 1970)

TABLE 2

CLAIMS TO NATIONAL JURISDICTION FOR NINE COUNTRIES BORDERING THE NORTH PACIFIC AS OF OCTOBER 1970
(In Nautical Miles)

Countries	Territorial Sea	Exclusive Fishery Zone*	Total*
United States	3	9	12
Canada	12	—	12
Mexico	12	—	12
USSR	12	—	12
Japan	3	—	3
South Korea	(not clear)	(not clear)	(not clear)
North Korea	12 (presumably)	—	12 (presumably)
China (Taiwan)	3	—	3
China (Mainland)	12	—†	12

* Not including areas specified under bilateral agreements (such as the Japan-South Korea agreement).
† Certain areas are closed to fishing.

Criticisms of the Existing Arrangements

The above-outlined regime for international regulation of fisheries is quite imperfect, and has been severely criticized for its inefficiency and lack of principles. Let us examine some of the criticisms.

It has been pointed out rightly that the power vested in international fishery bodies is limited. For example, most of the fishery commissions are authorized only to make recommendations to the member governments concerning regulatory measures. There is no guarantee that recommendations based on scientific evidence will be accepted. In practice, most of the recommendations are accepted because they represent compromises made through negotiations between the national delegations, but this is beside the point. Generally, they also lack the ability to cope with new problems speedily. Some of the member nations may demand watertight scientific evidence, which is almost impossible to obtain. Delegation of authority, however, is basically up to the member states. An international body, bilateral, regional, or global, can only be effective to the extent that it is empowered by the member states. From the other side, there is nothing an international fishery body cannot do if the member states agree.

The international fishery bodies are often criticized for their inability to prevent nonmember nations from entering the fisheries concerned and upsetting the existing arrangements. There is no real solution to the problem. No matter what sort of an international body nations may establish, it has to face the question of what to do with new countries entering the fisheries it regulates. Some conventions are open to new entry, while others are not. Under an open convention, such newcomers may become members and, if the total catch is fixed, allowance will have to be made for their shares of it, thus reducing the shares of the existing members. If the convention is closed, the problem must be handled by other means. It has been suggested that transferable property rights be established for a particular fishery so that only a limited number of operating units (or nations) can fish by obtaining licenses (which may be sold or leased to others) from the international authority managing the fishery. Aside from complicated administrative and political problems which would arise from such a system, I know of no country prepared to accept this as a common principle of international fishery regulation. In most nations, such a system has not been established even for fisheries under the jurisdiction of their national authorities.

Critics also say that most of the existing international arrangements for high seas fisheries have little value from an economic point of view since their goal is merely to maximize physical yields from certain stocks or areas, which may be quite different from economically most desirable yields (e.g., Christy and Scott, 1965; Crutchfield, 1970). This, too, has been used as a reason for suggesting such an international licensing system as mentioned above. Most of the people experienced in practical problems of international fishery agreements, however, do not consider that the maximization of physical yields is an overall solution; nor is it true that this criterion has been used as a basis for the international management of fisheries because biologists have largely been responsible for developing the existing arrangements. The concept of maximizing physical yields has generally been adopted, mainly because it has been more readily acceptable, as a common denominator, to the parties concerned than any other criterion. As we all know, even this goal is not easy to achieve except under a relatively simple set of conditions. It is also untrue that principles other than the maximization of physical yields have not been applied. On the contrary, a number of existing agreements are dealing intensively with the allocation of resources or catch, as illustrated in the first section of this paper.

Another criticism is concerned with the limited number of international fishery bodies employing an independent research staff. Only three fishery commissions conduct their own research using a full-time staff: the International Pacific Halibut Commission, the International Pacific Salmon Fisheries Commission, and the Inter-American Tropical Tuna Commission. The first two operate under conventions between the United States and Canada, and the third was established originally under a convention between the United States and Costa Rica, other parties joining later. It is generally considered preferable for any international fishery body to have its own research staff rather than depend on research conducted by various agencies of the member states, for this helps maintain objectivity and consistency in the planning, execution, and evaluation of research programs. Most of the nations in the world, however, are reluctant to meet expenditures to support an independent international research staff. Even the United States has not been increasing, in recent years, financial support to some of the commissions to which it is a party, while the amount of research work required has been growing. A general trend is for newly formed fishery bodies, which often include less-developed countries, to depend on research carried out by national agencies.

Research conducted by national agencies under a coordinated program may even be preferable under certain circumstances. For example, when the member states of a commission have varying interest in the different resources it manages, as is the case with the ICNAF or the Northeast Atlantic Fisheries Commission (NEAFC) this may be the only practical way of organizing research to cover most of the important resources. Many ambitious research programs have actually been conducted by coordinating the research efforts of national agencies under the auspices of international bodies, such as the ICES, the ICNAF, and the International North Pacific Fisheries Commission.

Arrangements for enforcing regulations are also criticized. Many treaties leave the power to enforce international fishery regulations to member states as far as their own flag vessels are concerned, thus making policing ineffective. This has been corrected gradually, and a few fishery agreements do provide for international inspection systems.

A major problem from the management point of view is the rapid rate at which a new large fishery may develop. As has been demonstrated by the development and expansion of fishing activities in the North Pacific, the Northwest Atlantic, the west coast of Africa, the west coast of South America, and elsewhere, the exploitation of a fishery resource can easily reach its maximum level within a matter of a few years. Thus, management problems often develop much faster than concrete evidence, on the basis of which regulatory measures are to be recommended, is obtained from time-consuming investigations. In some instances, changes in catch and CPUE (catch per unit effort) are so drastic that everybody agrees to take immediate measures. Generally speaking, too, fishing nations are increasingly aware of the problem; in fact, from the late sixties on, some major decisions have been made speedily, sometimes on the basis of rather inconclusive evidence. International fishery negotiations are improving, albeit slowly, in this regard.

A number of people have suggested that a global international agency be established to control all high seas fishing. Fishery problems are difficult enough to handle even on a regional basis. Difficulty will be multiplied by globalizing problems, for a vastly increased number of sovereign states, many of which have little or no direct interest in the fisheries concerned, would have a voice on everything. The degree of delegation of authority to such a global agency would be minimal, as evidenced by the past performance of the United Nations and its agencies. For example, agreement by at most seven or eight countries would be needed to cope with any fishery problem which might arise in the North Pacific in the foreseeable future. I see no reason for involving 120 more nations with greatly diversified interest. It will be helpful, however, to have a global agency to monitor the conditions of high seas fisheries and facilitate the conclusion of international agreements for their management, a function at present performed to a degree by FAO.

In short, as general principles of the law of the sea have never been adequate to handle complicated international fishery problems, practical arrangements of wide variety have been developed for international regulation of fishing activities under different principles and management practices. Many of them cover broad areas of the high seas, often one half of a large ocean such as the eastern North Pacific, the western North Pacific, the eastern North Atlantic, the western North Atlantic, and the tropical eastern Pacific. The new Atlantic tuna convention covers practically all waters of the Atlantic where tunas occur; the whaling convention is concerned with the whale resources of the entire world ocean. In addition, there exist a number of bilateral agreements of a more limited scope. We cannot realistically expect these various arrangements to be consolidated in a single system operating under a widely accepted set of principles. The international aspects of the utilization of living marine resources cannot be dealt with as new problems to which a simple and overall solution is to be found. New principles or patterns may evolve from the existing arrangements to meet new situations arising from further development of world fisheries.

Current Issues

One characteristic of the recent debates on ocean affairs is that different aspects of use of the ocean tend to be considered together. From a technical and legal point of view, international problems of fisheries can better be handled separately, for example, from those of mineral resources. On the international political scene, however, there is a growing tendency for all issues of the ocean to be considered to form a package deal.

Ocean resources come under two major categories, the extractive resources and nonextractive resources. Water, minerals and other chemicals, as well as living resources, which are removed from the sea for man's use, constitute the former. Other uses of the sea, including shipping and navigation, communication, recreational uses, waste disposal, military uses, and so on, which do not involve extraction of substances from the sea, can be called nonextractive resources (Schaefer, 1970). Among the extractive resources of the sea, the values of fish and oil (and gas) far exceed those of all others. As far as mineral resources are concerned, nothing is likely to come close to oil and gas in economic importance, at least in the foreseeable future.

In many ways, some of the nonextractive uses are far more important than the extractive resources of the sea. Among them are shipping, pursuit of recreation, military uses, and perhaps waste disposal. Revenues directly derived from shipping and related businesses far exceed the combined value of products from all extractive resources of the sea. Furthermore, the economies of most nations depend, to a large extent, and in some cases almost entirely, on seaborne trade. It is difficult to evaluate, in dollar value, the importance of recreational uses of the sea, but in affluent nations the general public tends to consider them more important than commercial fishing or even oil drilling. The value of military uses of the ocean is also not reducible to economic terms. For some nations, or under certain circumstances, military considerations supersede all other interests. Use of the ocean for waste disposal, too, cannot be assessed in monetary terms. With its enormous absorbing capacity, the ocean is playing an increasingly important role in this essential aspect of man's activity.

The last Law of the Sea Conference sponsored by the United Nations was held in 1958 and the last of the conventions resulting from the conference entered into force only in 1966. Yet a new round of discussions began in 1966 at the General Assembly and its committees. At the beginning emphasis was on surveys of what was known about various aspects of the ocean and its resources.

Political issues were brought in with a proposal by the delegation of Malta, which resulted in the establishment of an *ad hoc* committee to study the topic introduced, that is, "examination of the question of the reservation exclusively for peaceful purposes of the sea-bed and ocean floor, and the subsoil thereof, underlying the high seas beyond the limits of present national jurisdiction, and the use of their resources in the interest of mankind." The *ad hoc* committee was later replaced with a permanent committee composed of forty-two member states.

While the committee was focusing its attention on the question of a possible international sea-bed regime, the General Assembly, in 1970, instructed the secretary-general to ascertain the views of member states on the desirability of convening at an early date a conference on the law of the sea to review not only the sea-bed issue but also the regimes of the high seas, the continental shelf, the territorial sea and contiguous zone, fishing and conservation of the living resources of the high seas, and so on. The Conference of the Committee on Disarmament took up the question of sea-bed arms control, and the United States and the USSR worked out a joint draft, which has been revised for endorsement by the General Assembly.

The president of the United States issued, on May 23, 1970, a statement on United States ocean policy, which outlined proposals for a sea-bed treaty. Among the main features of the proposed treaty were the renouncement of all national claims over the natural resources of the sea bed beyond a depth of two hundred meters; the establishment of an international trusteeship zone comprised of the continental margins to be administered by each coastal state acting as trustee; and the establishment of international machinery to authorize and regulate use of sea-bed resources beyond the continental margins. The statement also indicated the desire of the United States to negotiate a new treaty which would establish a twelve-mile limit for territorial seas and provide for free transit through international straits. It would also accommodate the problems of developing countries and other nations regarding the conservation and use of the living resources of the high seas.

A draft sea-bed treaty prepared by the United States was submitted to a session of the Seabed Committee held in August, 1970. The Soviet and Latin American nations strongly objected to the draft. Many others felt that this formed a basis for further discussion but did not take any position at that time.

The United Nations and its specialized agencies, as well as other national and international institutions, have been debating on the international actions required for the prevention of marine pollution. Meanwhile, Canada has taken a unilateral action to establish a one-hundred-mile pollution control zone in the Arctic waters (an act to prevent pollution of areas of the waters adjacent to the mainland and islands of the Canadian Arctic, 1970).

In December 1970 a resolution was passed by the General Assembly to indicate a timetable for the next Law of the Sea Conference and the subjects to be discussed. The Seabed Committee was expanded to a membership of eighty-six and the conference was scheduled for 1973, subject to possible postponement. The committee would meet twice in 1971 to carry out preparatory work. The conference

> . . . would deal with the establishment of an equitable régime—including an international machinery—for the area and the resources of the sea-bed and the

ocean floor, and the subsoil thereof, beyond the limits of national jurisdiction, a precise definition of the area, and a broad range of related issues including those concerning the régimes of the high-seas, the continental shelf, the territorial sea (including the question of its breadth and the question of international straits) and contiguous zone, fishing and conservation of the living resources of the high seas (including the question of the preferential rights of coastal States), the preservation of the marine environment (including *inter alia,* the prevention of pollution), and scientific research; . . .

In short, the conference will not only take up new issues concerning the ocean but will also reopen all the issues debated at the 1958 conference.

Predictions

It is too early to predict what general principles of the law of the sea might emerge from the next conference. The relative importance of the subjects to be taken up differs greatly among the nations. For the United States, for example, a narrow territorial limit with free transit through international straits may be of utmost importance for defense reasons. A number of technologically advanced nations might take a similar position to minimize potential hazards to seaborne trade. A certain number of states, particularly those in Latin America, will adhere to their claim to extensive national jurisdiction not only to protect their interests in fisheries and other resources but also as a matter of national prestige. Others will perhaps try to justify their claims to national jurisdiction of one sort or another on the ground of special circumstances.

The political situation of the United Nations has changed radically since 1958. There will be some 130 votes at the next conference instead of 86 as in 1958. If the conference is held under the same rules as in 1958, and if all member states participate in voting for every issue, 87 votes will be required to adopt a proposal and 44 to block it. The large increase in the United Nations membership is accounted for almost entirely by less-developed countries which have become independent since World War II. These countries, the so-called developing nations, have become much more organized in making demands and blocking proposals they do not like, often against the wishes of technologically advanced nations (Burke, 1970).

Considering all the problems the conference is likely to face, there might not be a two thirds majority on any of the major issues. It is also possible that a treaty might be adopted by a two thirds majority, but without the participation of the bulk of the nations most concerned with the issue involved, as was the case with the fishing and conservation treaty of 1958. There is also a good chance that the conference may be further postponed because of lack of agreement at the preparatory meetings.

Under most favorable conditions, a fishery treaty might result from the conference. Such a treaty would perhaps include only very general principles, or rather guidelines, under which specific arrangements might be developed. If the draft articles of such a treaty contain very specific formulas, they will not receive broad support. Looking at most recent developments in fishery negotiations, one can speculate on the guidelines which might be considered for inclusion in such a treaty. Among them are an obligation to take international conservation measures; the need for special considerations to protect coastal fisheries; the acceptance of the allocation of catch as a principle of international regulation of fisheries without spelling out details to implement it (such details should be left to bilateral or multilateral arrangements between the countries concerned); the treatment of anadromous species, particularly salmon, as a special case; the facilitation of international inspection systems; and the setting up of a group of specialists to make recommendations, upon request by a regional fishery body or the governments of individual states, on questions of a scientific nature. The task of the group of specialists would not be arbitration of political issues, but would be restricted to passing judgment, based on available evidence, on such questions as effects of fishing, the level of maximum sustainable yield, and the distribution and migration of particular populations.

In any case, fishery matters would also be considered in the context of such other issues as the extent and nature of territorial seas and exclusive fishery zones, the regime of the high seas, and the definition of living resources subject to the new sea-bed treaty. Fishery problems might be used by some nations as trade-offs to gain ground on other issues.

Regardless of the outcome of the new Law of the Sea Conference, however, some predictions can be made on future trends as far as fishery problems are concerned. There will be a continuing trend for extension of national jurisdiction. It might take the form of broader territorial seas, or establishment or extension of exclusive fishery zones, or preferential rights of coastal states. National claims might also be expanded through a new definition of living resources subject to the existing continental shelf convention and/or a new sea-bed treaty. It is also possible that some nations might translate the new regime for sea-bed resources into a regime for the control of living resources in superjacent waters. Whether a nation likes it or not, the world is moving toward extended national jurisdiction over the exploitation of living resources.

In parallel with the further extension of national jurisdiction, there will be a trend for more bilateral and multilateral arrangements for specific fisheries and areas, perhaps with greater emphasis on the allocation of the catch from heavily exploited stocks. There will be many trade-offs, particularly in the case of bilateral agreements. The continuing trend for broader national claims will not obviate the need for specific international arrangements. Let us consider, for example, the real situation in the North Pacific. The country most likely to extend national jurisdiction is Canada. The United States might also take action to extend jurisdiction over some of the resources now fished mainly by foreign fleets. But this would not eliminate the need for bilateral or multilateral arrangements. The United States and Canada would have to continue the Fraser River salmon treaty, and it is unlikely that they would wish to terminate the halibut treaty to divide up halibut fishing grounds off their coasts. The fur seal treaty would continue in any case. I am sure that there would still be some arrangement to protect the Canadian and American salmon stocks. The USSR and Japan are at present not likely to extend national jurisdiction very far; the agreements between them would remain. A substantial part of the tropical tuna fishing grounds would still be beyond the limits of national jurisdiction, with international grounds expanding further offshore; the present convention or an arrangement of a similar nature would continue to be required.

Extension of national jurisdiction will perhaps result in an even more complicated network of international agreements in some parts of the world. As a remote possibility, one can think of the trouble of establishing permanent national boundaries for all waters within two hundred miles of the sovereign states and territories in Southeast Asia, the Caribbean, much of Europe, or the South Pacific; it might take hundreds of years to settle some of the conflicts arising from this aspect alone. Special problems of fisheries could also be quite complex and fluid. Each of a number of countries along the West African coast, for example, has a rather short stretch of coastline. Should all of them extend national jurisdiction very far, their own fishing activities would be restricted greatly. They would at least have to make some sort of regional arrangement among themselves. In Southeast Asia, various nations are utilizing the resources of the South China Sea and adjacent waters. A chaotic situation would arise if they should extend national jurisdiction in all directions without making a regional arrangement.

October 1970

REFERENCES

Burke, W. 1970. "Law, Science, and the Ocean." *Natural Resources Lawyer*, 3 (No. 2): 195–226.

Christy, F. T., Jr., and A. Scott. 1965. *The Common Wealth in Ocean Fisheries*. Baltimore, Md.: The Johns Hopkins Press.

Crutchfield, J. A. 1970. "Economic Aspects of International Fishing Conventions." In A. D. Scott (ed.), *Economics of Fisheries Management,* pp. 63–77. H. R. Macmillan Lectures in Fisheries. Vancouver: University of British Columbia, Institute of Animal Resources Ecology.

Harlow, B. A. 1968. "Legal Aspects of Claims to Jurisdiction in Coastal Waters." In DeWitt Gilbert (ed.), *Future of the Fishing Industry of the United States,* pp. 310–20. University of Washington Publication in Fisheries, n.s. 4.

Kasahara, H. 1961. *Fisheries Resources of the North Pacific Ocean, Part I.* H. R. Macmillan Lectures in Fisheries. Vancouver: University of British Columbia.

———. 1965. "Japanese Fisheries and Fishery Regulations." *California and the World Ocean,* pp. 57–66. California Museum of Science and Industry.

———. 1970. "International Aspects of the Exploitation of the Living Resources of the Sea." Paper presented at a preparatory conference (February 1970, New York) of *Pacem in Maribus,* Malta, 1970.

Leonard, L. L. 1944. *International Regulation of Fisheries.* Carnegie Endowment for International Peace, Division of International Law, Monograph No. 7. Concord, N.H.: Rumford Press.

Schaefer, M. B. 1970. "The Resources Base and Prospective Rates of Development in Relation to Planning Requirements for a Regime for Ocean Resources beyond the Limits of National Jurisdiction." Paper presented at a preparatory conference (February 1970, New York) of *Pacem in Maribus,* Malta, 1970.

3

World Fisheries—World Concern

DONALD L. McKERNAN

In the sixties fisheries production throughout the world continued to increase at a healthy pace. The interest that world fisheries have engendered among governments and people has increased even more as evidenced by statements made by many governments at preparatory meetings for the Law of the Sea Conference scheduled for 1973 by the United Nations. The world fish catch reached 64.3 millions of metric tons in 1968, but in the last year or two the catch has declined to about 63 million tons. The annual rate of increase in fish catch until 1968 had been between 5 and 6 percent but now with a declining rate the total catch has declined.*

Governments and fishing industries have not yet reacted to this decline and for the most part the industries of both the maritime nations such as the Soviet Union, Japan, and most of the Western European nations on one hand, and on the other hand the governments of many developing coastal nations such as Senegal, Ghana, Brazil, Chile, and Peru have continued to react as though the potential world fish catch, and particularly the catch in those areas of interest and concern to them, could continue to increase almost without limit. It is true that some of the optimistic estimates of the potential world fish catch made by experts during the late sixties could lead fishermen to believe that the fishery potentials of the world ocean have indeed not yet been reached, as evidenced by the estimates presented in Table 3. Dr. W. M. Chapman (1966), for example, estimated that most likely a thousand million metric tons of fish could be harvested from the world ocean. Of course there have been some more conservative estimates but these have not received the wide

Donald L. McKernan is presently coordinator of ocean affairs in the Department of State and special assistant to the secretary of state for fisheries and wildlife, with the rank of ambassador. He is responsible for supervising United States activities in nine active international fisheries commissions, as well as those associated with United Nations organizations dealing with fisheries, wildlife, and oceanography. He was formerly director of the Bureau of Commercial Fisheries, administrator of commercial fisheries in Alaska for the United States Fish and Wildlife Service, and assistant director of the Pacific Oceanic Fishery Investigations.

* The 1970 world fish catch again rose to a new high of 69 million tons, but it seems doubtful if that level of increase can be maintained.

TABLE 3

ESTIMATES OF TOTAL OCEAN YIELDS OF AQUATIC ANIMALS

Forecast (million metric tons)	Method *	Author
50 to 60	ext.	Finn (1960)
55 (bony fishes)	ext.	Graham and Edwards (1962)
55 (by 1970)	ext.	Meseck (1962)
60 (bony fishes)	ext. mf.	Graham and Edwards (1962)
66 (by 1970)	ext.	Schaefer (1965)
70 (by 1980)	ext.	Meseck (1962)
60 to 80	ext.	Alverson and Schaefers (1965
70 to 80	ext. mf.	Bogdanov (1965)
115 (bony fishes)	mf.	Graham and Edwards (1962)
160	ext.	Schaefer (1965)
200	mf.	Schaefer (1965)
200	mf.	Pike and Spilhaus (1965)
1,000	mf.	Chapman (1966)
180–1,400	mf.	Pike and Spilhaus (1962)
2,000	mf.	Chapman (1965)

Source: Schaefer and Alverson (1968:82).
* ext.: extrapolated from catch trends or existing knowledge of world fish resources; mf.: energy flow through food chain.

publicity of the more optimistic forecasts, and many nations looking at the world demand for food turn to the sea, optimistically, to supply this demand. It seems useful to point out some of the factors that have been involved in recent world fisheries developments; some of the constraints that are preventing further increases; and some of the problems facing the world fishing industry as we enter a decade of debate on the rules governing ocean fishing.

Conflicts between fishermen of different nations are increasing, and the views of the nations as to how these conflicts should be resolved promise an active period of debate and negotiation as the community of nations seeks to bring some order out of the current chaos. The very rapid expansion of world fish catch, which has tripled the yield since 1945, has resulted in the overfishing of many desirable stocks of coastal fish and has brought an awareness to many nations that the current rules (or lack of rules) by which conflicts are resolved between fishermen of various nations are inadequate for the coming decades. Such conflicts affect broadly the foreign relations of countries; their security as it depends upon the oceans; the freedom of research; the preservation of the marine environment; and, in fact, for all practical purposes all uses of the sea.

In view of these conflicts and in view of the importance of arriving at an appropriate means of resolving fishing disputes there are a few observations which can be made at the start:

1. The recent annual increase in the rate of production of world fish catch of between 5 and 6 percent cannot be maintained throughout the decade of the seventies.

2. In spite of the views of many coastal fishermen, there are places in the world where fishery resources are not fully utilized, and although the cost of catching fish in these areas and the cost of marketing them are more difficult and more expensive, the world fish catch can continue to increase. For example, the Indian Ocean can produce much more fish, perhaps as much as three or four times its current production. The east and west coast of Africa will produce additional tonnages of fish. The Patagonian shelf off the east coast of South America has not been heavily fished at present and can yield many times its present catch. The Antarctic will provide an increase in fish catch, and if greater use can be made of krill, the crustacean in great abundance in Antarctic waters (which also forms the major food of the Antarctic baleen whale populations), the catch in the southern hemisphere will substantially increase. One can predict that the world tuna catch will be increased as fishermen's ability to catch the surface tuna schools in the central western Pacific and Indian Ocean improve. A number of pelagic species such as saury, mackerel, and other pelagic species will undoubtedly contribute to further increases in the world fish catch.

3. An increase in world catch can only occur if conservation of presently used resources is improved. It seems increasingly evident, even in the past few years (McKernan, 1968), that current international, national, and state conservation efforts are not successfully preventing the depletion of fishery resources of the greatest economic importance.

4. The present rules governing fishing activities are not only unsatisfactory from the standpoint of conserving the resources, that is, preventing depletion, but they seem incapable of providing a basis for equitable allocation of these resources among fishermen or nations.

5. Present practices of fishing throughout the world favor maritime countries with large, highly mobile fishing fleets and discriminate against coastal fisheries using smaller, less mobile vessels. These practices encourage the overconstruction of large mobile fishing fleets; they encourage the so-called "pulse fishing"; and they discourage the rational use and stable economic development of fisheries.

6. The present conservation arrangements within the United States are for the most part ineffective in conserving important fishery resources controlled entirely by the United States.

7. Neither the twelve-mile fishery jurisdiction of the United States nor the two-hundred-mile fishery jurisdiction of several Latin American coun-

tries guarantees effective conservation any more than the multilateral international fishery conventions now in effect.

Let us examine in greater detail the contributing factors which will affect future development of world fisheries. As stated above, evidence is mounting that the recent rate of increase in production of world fish catch cannot continue. This is so primarily because the application of fishery management techniques has not kept pace with the increases in fishing effort and we have failed to prevent the adverse effect of increased competition and increased effort on the limited available stocks of many coastal species of fish. Our experience throughout the world with such major fisheries as Pacific sardine, herring, cod, the Antarctic whale, the northern fur seal, salmon, haddock, menhaden, and others has proven that aquatic animals, whether they be whales or menhaden, are not inexhaustible, and it has become evident that we can overharvest them. Even in the case where some of these stocks are not overfished in a technical sense we have so overcapitalized and overbuilt the fishing fleets that it has become virtually uneconomic to pursue the fishery. It seems quite clear that the international rules by which coastal and high seas fisheries operate must be re-ordered to prevent overfishing and the uneconomic use of the heavily fished species. Until this is done, it would appear that the continued overfishing of many important stocks will so reduce the catch of these species (e.g., herring, cod, menhaden, etc.) that the decline of the most heavily fished species will balance the increase in production from newly exploited world fishery resources and world fish catch will remain about level.

There are places in the world ocean where the stocks of fish have not been heavily fished. Stocks of herring, saury, shrimp, and certain ground fish are underfished in the eastern North Pacific, and it seems obvious that there are underharvested resources in many other parts of the world ocean. If the conservation question as well as the allocation of coastal resources can be resolved, it seems probable that world fish catch will again increase and that this increase will come from the Pacific, Indian, and Southern oceans.

Bilateral and multilateral conservation conventions have been quite successful in the past few decades in conserving important fishery resources, but it is obvious that these same methods are becoming less successful as we enter the seventies. Evidence is accumulating that the present national and international conservation arrangements, including such classic conventions as the International Pacific Halibut Convention with Canada and the International North Pacific Fisheries Convention with Canada and Japan, are out of date and do not take into account the capability of modern fishing fleets or the tremendous efficiency of modern fishing gear to deplete the resources. We must improve the performance of the commissions which have resulted from such conventions or find new ways to provide rational control of world fisheries. For example, currently applied conservation methods do not adequately take into account the economic interests of certain of the fishermen using fish stocks under the control of these commissions. They do not provide for a reasonable system of allocation of these resources among governments and fishermen, and thus lead to increased competition and increased fishing effort for the limited available catches. This result has led to a reduced income per unit of effort and a waste of manpower and capital as well as to an increased threat to the conservation of the resources.

Present practices under current law of the sea favor the maritime countries and discriminate against coastal nations—a common complaint by coastal fishermen. An examination of the coastal fisheries of the United States shows good examples of this discrimination. Take for instance the appearance of more than two hundred foreign fishing vessels, most of which are of very modern design and very large by this country's standards, fishing off the east coast of the United States. These vessels in the early 1960s began fishing the limited haddock stocks found on Georges Bank. This foreign fishing, coupled with the traditional and intensive fishing by United States and Canadian fishermen, led to the depletion of the stocks of haddock. Following this reduction of stock there has been a drastic reduction in annual recruitment of young haddock. The current sustainable yield from Georges Bank haddock stock is at a very low level and scientists recommend that no commercial fishing take place on this stock for several years. The heavy influx of foreign fishing, which occurs without regard to the effects of such an increase on limited local stocks of fish, often causes not only the depletion of the resources in question but also severe economic loss to the affected coastal fishermen. American fishermen who have fished this stock of haddock for over a hundred years have been seriously affected, and have few alternative sources of supply to make up for the loss of haddock. In the meantime, the foreign fleets have moved on to other areas and other species such as the herring and mackerel stocks of Georges Bank. Such practices, whether by foreign or United States fishermen, will not be tolerated by fishermen in the future and the recent reactions of coastal nations show this quite clearly.

It was previously mentioned that the conservation methods practiced within the United States are equally ineffective in conserving purely domestic resources. Witness the current depletion of United States menhaden on the Atlantic coast. The Atlantic coast catch of menhaden has declined from 648,000 short tons in 1961 to 196,000 in 1969 and 291,000 in 1971. On the Gulf coast the menhaden catch has fluctuated but has not as yet shown signs of overfishing, although recent high catches and effort do not give one confidence that this stock can maintain the current levels of catch and effort for long (Table 4). The Pacific sardine is still another example of overfishing and poor conservation of a strictly national resource (Murphy, 1966). The same can be said for Atlantic oysters, Alaskan king crab, and a number of other important resources.

Our current system of conservation which provides for control by local political units—states and even countries—brings about political regulation of fisheries rather than technical and scientific management and prevents the rational control of the fishing effort and catch. It also adversely affects the competitive position of the fishermen in relation to foreign fishermen and, for the most part, is a major factor why an otherwise efficient United States fisherman is among the most inefficient in the world.

Turning again to the international scene, since World War II there have been important developments in world fisheries, and many of these have affected the catch of fish off the coast of the United States and United States fisheries. It would not be incorrect to say that the postwar fisheries developments off the coast of the United States—involving the appearance for the first time of fishing fleets of about twelve nations—have been as varied and complicated in terms of multinational competition for limited resources as has occurred anywhere in the world. In the eastern North Pacific Ocean, the fisheries development since 1945 has been meteoric. The prewar fisheries of Alaska were almost entirely for salmon, and by United States citizens, although in the earlier years there was a small United States fishery for herring, dungeness crab, and shrimp. After World War II, the United States fisheries for these species and for king and tanner crab increased dramatically, and both Japan and the Soviet Union entered this fishery and have increased their catch from waters off Alaska many-fold during the fifties and sixties. It is probable that the total catch of all the species from the eastern North Pacific Ocean at the present time is at least two million metric tons and may in fact be approaching three million tons and includes very large fisheries for Alaskan pollack, ocean perch, and several species of flounders, as well as less intense fisheries for black cod, halibut, salmon, shrimp, and king and tanner crab.

A recent development has been the initiation of a large Soviet fishery off the coast of Washington, Oregon, and northern California for Pacific hake. The Soviet catch has been in the neighborhood of 150,000 tons of hake per year in recent years, and the appearance of a large Soviet fleet in an area where small-craft operations from United States ports are very heavy has brought into sharp focus the problem arising from the conflict between coastal fishing vessels and the large distant-water vessels of the Soviet Union. The three active fisheries commissions in this area are not constituted to cope with the problems arising from the recent rapid

TABLE 4

UNITED STATES MENHADEN CATCH: 1961–70
(Short Tons)

	Atlantic Coast	Gulf Coast	Total
1961	648,000	510,000	1,158,000
1962	646,000	527,000	1,173,000
1963	424,000	484,000	908,000
1964	333,000	452,000	785,000
1965	352,000	511,000	863,000
1966	258,000	396,000	654,000
1967	232,000	350,000	582,000
1968	277,000	413,000	690,000
1969	196,000	578,000	774,000
1970	314,013	604,733	918,746
1971	291,597	804,439	1,095,036

increase in fisheries in the eastern North Pacific and the hundreds of foreign fishing vessels of Japan, the Soviet Union, and most recently South Korea.

For example, the International Pacific Halibut Convention between the United States and Canada has been uniquely successful during the past forty years, rehabilitating the halibut stocks which had been depleted by the fishermen of the two countries in the early twentieth century and increasing the level of sustainable yield to the maximum at about seventy million pounds per year. Since the advent of the very large bottom fish fisheries in the Bering Sea and North Pacific by Japan and the Soviet Union, and most recently by South Korea, it seems quite obvious that the sustainable yield of halibut available to the United States and Canada is decreasing. It is likely that it will continue to decrease even though it is not yet possible to prove the deleterious effect the trawl fisheries of the two distant-water nations are having on the productivity of the eastern North Pacific halibut stock. By the same token, the International North Pacific Fisheries Convention has been successful in providing a basis for rehabilitation of many of the salmon stocks of the eastern North Pacific and furthermore has provided a reasonable basis for allocation of the allowable salmon catch to Japan, the United States, and Canada.

It is true that on one hand Japan is unhappy with the "provisional abstention line" at 175 degrees west longitude which prohibits the Japanese salmon fishermen from fishing close to the United States and Canadian coasts in the eastern North Pacific and, on the other hand, the United States and Canada are not satisfied with the approximately three million salmon of North American origin taken by the Japanese high seas fisheries each year. Still, the present convention has enabled United

States and Canadian authorities to apply effective regulations against the inshore fisheries of the two nations and the major stocks of salmon of both countries appear to be on the increase. Recent indications are that both South Korea and the Soviet Union may be looking at the salmon stocks of the North Pacific. The Soviet Union's negotiations with Japan in 1971 regarding the high seas salmon catch in the western North Pacific provided for the initiation of a high seas salmon fishery by the Soviet Union. In both 1969 and 1970 South Korean fishermen took several thousand salmon of North American origin causing great concern among the fishermen of the United States and Canada. The INPFC has not provided effective conservation for ground fish species other than a measure of protection for halibut, and United States authorities believe that many stocks of ground fish, including the Pacific ocean perch, several species of flounders, and Alaska pollack are probably being overfished in the Bering Sea by the Soviet Union and Japan.

The International Pacific Salmon Fisheries Commission has successfully regulated the sockeye salmon runs of the Fraser River and Puget Sound and divided the catch between Canada and the United States for many years. In recent years, this commission has also been quite successful in providing improved conservation for the pink salmon runs of the same area. However, the rapid increase in the troll fishery for salmon in the Pacific Northwest and Alaska has now convinced both Canada and the United States that the present convention is inadequate to cope with the current conditions in the salmon fisheries. Both nations are discussing an expansion of the convention which, among other things, will provide for a broader allocation of the stocks among the fishermen of the two nations. In summary, the ten-fold increase in fisheries in the eastern North Pacific following World War II has brought new problems of conservation of fisheries resources to the area and has caused increasing conflicts between the fishermen of those nations bordering the North Pacific and fishing common stocks of fish.

The very rapid increase in the tuna catch in the eastern tropical Pacific following World War II led to a 1949 convention between the United States and Costa Rica and the formation of the Inter-American Tropical Tuna Commission. Since that time, the tuna catch in this area has increased substantially and four additional nations, Panama, Mexico, Canada, and Japan, have joined the commission. The commission hired an exceptionally competent staff led by the late Dr. Milner B. Schaefer and began studying the tuna resources under exploitation by member nations. In 1966 the commission first adopted conservation regulations limiting the catch of yellowfin tuna in the area. Since that time, increasingly stringent regulations have been applied, especially in view of the increased fishing effort, and the season for yellowfin tuna has been reduced substantially since the inception of regulations. The developing nations who are members of the commission are not satisfied with the present regulatory system and the threat of additional fleets of tuna vessels from nations located outside the area as well as significant increases in fishing effort from the United States and other member nations threaten the continued success of the commission at this time. Some member nations insist that the allowable catch of tuna be allocated among member states with special preferences to developing countries and to countries off whose coast the tuna are taken. This difference in view has led to the establishment of a working group within the member states, and studies continue on the effects of the present regulatory system on the interests of the member states with an examination of possible alternative regulatory systems which would more "equitably" divide the allowable catch. In spite of these differences between members, the convention has been uniquely successful in maintaining the tuna stocks at a very high level of productivity thus far.

The Gulf of Mexico and Caribbean are dominated by the great shrimp fishery developed by the United States, although as mentioned before the Gulf menhaden fishery is also an important industry in this region. The shrimp fishery of the Gulf of Mexico produced about 115,000 short tons in 1970 with a value of about $108 million. The total catch of shrimp in the United States in 1970 was approximately 183,000 tons valued at $130 million, by far the most valuable group of species in the United States fishery. At the same time, important segments of the United States fishing industry have established freezing and transportation facilities in other Gulf and Caribbean nations and a significant proportion of the imports of shrimp into the United States, amounting to a record importation of about 110,000 tons in 1970, were taken by United States flag vessels fishing off the coast of South America and out of Caribbean ports. As yet there is no active international convention regulating the catch of shrimp in the Gulf and Caribbean although this clearly is necessary in view of the very great demand and the limited supply of this resource. Several of the nations in the Caribbean and the nations bordering the western Atlantic have extended their jurisdiction over fisheries off their coast in an attempt to limit the foreign fishing. Conflicts between the United States fishermen and the fishermen and governments of several of these countries occur quite regularly and it seems essential that some international arrangement be initiated soon to provide for the investigation and conservation of the shrimp resources and eventually for some reasonable allocation of the catch among nations.

One of the greatest developments in fisheries has occurred in the North Atlantic where the developed nations of northern Europe, the Soviet Union, Canada, and the United States compete for about fifteen million metric tons of fish annually. Large numbers of vessels are competing for the limited resources available and both conservation conventions operating in this area, the International Convention for the Northwest Atlantic Fisheries and the Northeast Atlantic Fisheries Convention, have been unable in recent years to control fishing effort adequately. Herring, cod, haddock, and yellowtail flounder in the northwestern Atlantic are declining, some drastically. Total catches in the North Atlantic, which increased approximately 30 percent by 1967 over the 1963 level, have recently dropped 10 percent, with even greater declines in catches of herring in the northeastern Atlantic and haddock in the northwestern Atlantic. ICNAF has regulated certain of the fisheries of the northwestern Atlantic for many years by regulating the size of the mesh of the bottom trawls used for catching ground fish. In earlier years with much less fishing effort, scientists working with the commission believed the gear mesh regulations were effective in maintaining the stocks of cod and haddock at high levels of productivity. However, the enormous increase in effort which occurred during the 1960s has brought about the inevitable overfishing of many of the stocks and the decimation of some. The commission has reacted by providing for more rapid implementation of commission recommendations, by recommending provisions which will allow a division of the allowable catch among nations, and by introducing effective international enforcement into the convention waters. It remains to be seen whether this is "too little and too late." At the present time scientists and administrators closely watching the developments in the western North Atlantic fisheries are not overly optimistic that the present convention is an adequate mechanism for conserving the stocks of fish or providing a reasonable allocation of catch among the fifteen to twenty nations now fishing the area.

Similar rapid advances in fishing techniques and rapid increases in fishing effort by both coastal and distant-water nations are creating conservation and distribution problems throughout the world, to which nations are reacting in various ways. Two trends in world fisheries management seem to be developing simultaneously. Both arise from the view that the current institutions are inadequate to rationally conserve and equitably allocate the catch of world fishery resources. One trend would broaden the use of regional fisheries conservation and management conventions, attempt to improve the operations of these conventions, and seek ways within regional conventions to provide for better conservation and equitable allocation of allowable catches. Obviously the developed nations who are fishing in many parts of the world are the strongest advocates for this development. Many of the developing nations and coastal fishing nations do not see this as a viable alternative to the present somewhat inadequate international fishery conventions. Many of these coastal nations, particularly the developing nations, believe not only that current practices do not allow fisheries development among the less developed countries but that the present regional commissions prevent such development and tend to widen the gap in the use of living resources of the sea between the developed and developing nations. Some of these nations have reacted by unilaterally extending their jurisdiction over fisheries to extreme limits of two hundred miles (Table 5). This, they think, will provide for the effective conservation of the resources off their coast and will give these coastal states the opportunity to develop the use of these resources in their own good time for the benefit of their people. Obviously these points of view lead to direct conflict and a resolution is not in sight.

One can reasonably conclude that the present regional arrangements in fisheries have not been wholly successful, nor has the unilateral extension of jurisdiction by coastal states been sufficient to insure the adequate conservation of resources or the positive development and full use of the renewable resources found off their coasts. The regional conventions on one hand have been useful in many respects. They have improved the exchange of statistical and biological information between member countries; they have stimulated research on the resources under exploitation; and they have, in some cases, effectively regulated the catch of the resources. Some have been successful in achieving conservation; others have failed. Some few have provided for equitable allocation of the catches (e.g., the International Pacific Salmon Fisheries Convention and the North Pacific Fur Seal Convention) while most others, including the Northwest Atlantic convention, have failed thus far, in spite of repeated attempts, to provide for national allocations of the allowable catch. The coastal countries such as Ecuador, Peru, and Chile who have unilaterally extended jurisdiction over fisheries to two hundred miles have not avoided difficulties and confrontation between themselves as coastal states and the distant-water states, including the United States. On the other hand, Peru cites the anchovetta fishery as an example of the success of the concept of extended fisheries jurisdiction. Still, the remaining resources of Peru are not being developed very rapidly and this is true of Ecuador and Chile as well. Cooperation between these three nations and the developed nations in the fisheries field is almost nonexistent, and until the conflicts arising from their different juridical positions has been resolved there is not likely to be much cooperation or assistance in the fisheries field.

One of the hopes for the future lies in a successful law of the sea convention coming out of the Law of the Sea Conference now set for 1973. Preparations for this conference have begun and there is hope that the nations of the world can compromise so that law of the sea as it

TABLE 5

Breadth of Territorial Sea and Fishing Jurisdiction Claimed by Members of the United Nations System

Country	Territorial Sea	Fishing Limits	Other
Afghanistan	No coast		
Albania	12 miles	12 miles	
Algeria	12 miles	12 miles	
Argentina	200 miles	200 miles	Continental shelf—including sovereignty over superjacent waters
Australia	3 miles	12 miles	
Austria	No coast		
Barbados	3 miles	3 miles	
Belgium	3 miles	12 miles *	
Bolivia	No coast		
Botswana	No coast		
Brazil	200 miles	200 miles	
Bulgaria	12 miles	12 miles	
Burma	12 miles	12 miles	
Burundi	No coast		
Byelorussian SSR	No coast		
Cambodia	12 miles	12 miles	Continental shelf—to 50 meters including sovereignty over superjacent waters
Cameroon	18 miles	18 miles	
Canada	12 miles	12 miles	Fisheries closing lines
Central African Republic	No coast		
Ceylon	12 miles	12 miles	Plus right to establish 100-mile conservation zones
Chad	No coast		
Chile	3 miles	200 miles	
China, Republic of	3 miles	3 miles	
Colombia	12 miles	12 miles	
Congo (Brazzaville)	15 miles	15 miles	
Congo (Kinshasa)	3 miles	3 miles	
Costa Rica	3 miles		"Specialized competence" over living resources to 200 miles
Cuba	3 miles	3 miles	
Cyprus	12 miles	12 miles	
Czechoslovakia	No coast		
Dahomey	12 miles	12 miles	100-mile mineral exploitation limit
Denmark	3 miles	12 miles *	
Greenland	3 miles	12 miles	
Faeroe Islands	3 miles	12 miles	
Dominican Republic	6 miles	12 miles	
Ecuador	200 miles	200 miles	
El Salvador	200 miles	200 miles	
Equatorial Guinea	12 miles	12 miles	
Ethiopia	12 miles	12 miles	
Fiji	3 miles	3 miles	
Finland	4 miles	4 miles	
France	3 miles	12 miles *	
Polynesia	3 miles	12 miles	
Other overseas areas	3 miles	3 miles	
Gabon	25 miles	25 miles	
Gambia	12 miles	12 miles	
Germany, Fed. Rep. of	3 miles	†	
Ghana	12 miles	12 miles	Undefined protective areas may be proclaimed seaward of territorial sea, and up to 100 miles seaward of territorial sea may be proclaimed fishing conservation zone
Greece	6 miles	6 miles	
Guatemala	12 miles	12 miles	
Guinea	130 miles	130 miles	
Guyana	3 miles	3 miles	
Haiti	6 miles	6 miles	
Honduras	12 miles	12 miles	
Hungary	No coast		
Iceland	4 miles	12 miles	
India	12 miles	12 miles	Plus right to establish 100-mile conservation zone

Indonesia	12 miles	12 miles	Archipelago concept baselines
Iran	12 miles	12 miles	
Iraq	12 miles	12 miles	
Ireland	3 miles	12 miles *	
Israel	6 miles	6 miles	
Italy	6 miles	12 miles *	
Ivory Coast	6 miles	12 miles	
Jamaica	12 miles	12 miles	
Japan	3 miles	3 miles	
Jordan	3 miles	3 miles	
Kenya	12 miles	12 miles	
Korea, Rep. of	3 miles	20–200 miles	Continental shelf—including sovereignty over superjacent waters
Kuwait	12 miles	12 miles	
Laos	No coast		
Lebanon	N. A.	6 miles	
Lesotho	No coast		
Liberia	12 miles	12 miles	
Libya	12 miles	12 miles	
Luxembourg	No coast	†	
Madagascar	12 miles	12 miles	
Malawi	No coast		
Malaysia	12 miles	12 miles	
Maldives	2.75–55.0 miles	100–150 miles	(Modified archipelago concept—box around islands)
Mali	No coast		
Malta	3 miles	3 miles	
Mauritania	12 miles	12 miles	
Mauritius	12 miles	12 miles	
Mexico	12 miles	12 miles	
Monaco	12 miles	12 miles	
Mongolia	No coast		
Morocco	12 miles	12 miles	Exception: 6 miles for Strait of Gibraltar
Nauru	3 miles	12 miles	
Nepal	No coast		
Netherlands	3 miles	12 miles †	
New Zealand	3 miles	12 miles	
Nicaragua	3 miles	200 miles	Continental shelf—including sovereignty over superjacent waters
Niger	No coast		
Nigeria	30 miles	30 miles	
Norway	4 miles	12 miles	
Oman	3 miles	3 miles	
Pakistan	12 miles	12 miles	Plus right to establish 100-mile conservation zones
Panama	200 miles	200 miles	Continental shelf—including sovereignty over superjacent waters
Paraguay	No coast		
Peru	200 miles	200 miles	
Philippines	0–300 miles		Archipelago concept baselines; waters between these baselines and limits described in Treaty of Paris, Dec. 10, 1968; United States–Spain treaty of Nov. 7, 1900; and United States–United Kingdom treaty of Jan. 2, 1930, considered to be the territorial sea.
Poland	3 miles	12 miles	
Portugal	6 miles	12 miles *	
Romania	12 miles	12 miles	
Rwanda	No coast		
Saudi Arabia	12 miles	12 miles	
Senegal	12 miles	12 miles	Plus 6-mile contiguous zone
Sierra Leone	12 miles	12 miles	
Singapore	3 miles	3 miles	
Somalia	12 miles	12 miles	
South Africa	6 miles	12 miles	
Spain	6 miles	12 miles *	
Sudan	12 miles	12 miles	
Swaziland	No coast		
Sweden	4 miles	12 miles *	
Switzerland	No coast		
Syria	12 miles	12 miles	Plus 6-mile "necessary supervision zone"
Tanzania	12 miles	12 miles	
Thailand	12 miles	12 miles	
Togo	12 miles	12 miles	
Tonga	3 miles	3 miles	
Trinidad & Tobago	12 miles	12 miles	
Tunisia	6 miles	12 miles	Exclusive fisheries zone follows the 50-meter isobath for part of coast (max. 65 miles)

TABLE 5 (*Continued*)

Country	Territorial Sea	Fishing Limits
Turkey	6 miles	12 miles
Black Sea	12 miles	
Uganda	No coast	
Ukrainian SSR	12 miles	12 miles
United Arab Republic	12 miles	12 miles
United Kingdom	3 miles	12 miles *
Bahama Islands	3 miles	12 miles
Other overseas areas	3 miles	3 miles
USSR	12 miles	12 miles
United States	3 miles	12 miles
Upper Volta	No coast	
Uruguay	200 miles	200 miles
Vatican City	No coast	
Venezuela	12 miles	12 miles
Vietnam, Rep. of	3 miles	20 k. (10.8 miles)
Western Samoa	3 miles	3 miles
Yemen (Aden)	12 miles	12 miles
Yemen (Sana)	12 miles	12 miles
Yugoslavia	10 miles	10 miles
Zambia	No coast	

Note: Information is based on the synoptical tables concerning the breadth and juridical status of the territorial sea and adjacent zones prepared for the 1958 and 1959 Geneva Law of the Sea conferences, a synoptical table prepared by FAO in 1971, and additional information available to the Department of State. All mileage figures refer to nautical miles.

UNRECOGNIZED REGIMES: The following unverified information represents to the best of our knowledge the jurisdictional claims of certain regimes which are not members of the United Nations system. Their inclusion in this list is for information purposes only and in no way implies recognition by the United States of the validity of these regimes.

East Germany	3 miles	3 miles
North Korea	12 miles	12 miles
North Vietnam	12 miles	12 miles
Peoples Rep. of China	12 miles	12 miles

* Parties to the European Fisheries Convention which provides for the right to establish 3-mile exclusive fishing zone seaward of 3-mile territorial sea plus additional 6-mile fishing zone restricted to the convention nations.

† Signatories of the European Fisheries Convention.

pertains to fisheries can provide a basis for a successful international management scheme.

Looking at the real objectives of both the coastal and fishing nations on one hand and the developed and developing nations on the other, one can see emerging certain principles that if accepted would provide a basis for reconciling differences in points of view. These principles can be stated simply.

1. Adequate national and international organizations—primarily regional in nature—must be formed to conserve the resources and provide for a high sustainable level of yield. Such organizations must be available to all nations and provide expertise to the developing as well as the developed world.

2. There must be a better balance between the rights of the coastal state and those of the distant-water fishing state. That is, preference must be given to the coastal states over fishery resources lying off their coasts and associated with coastal waters.

3. Where disputes arise between nations over fisheries there must be adequate means of resolving these conflicts in a timely manner protecting the rights of both parties.

4. A stable world fishery regime must take into account not only the interests of the coastal states located near productive fishing grounds and the distant-water states with their large and efficient fishing fleets but also the remainder of mankind who also have an interest in the food resources of the world ocean, including landlocked and shelf-locked states and those states with narrow access to the sea who may not have productive seas adjacent to their coasts.

5. There must be some effective guidelines for the ultimate allocation of the resources among nations of the world.

6. There must be enough of an international regime to ensure accurate registration of the amount and changes in fishing effort, compilation of adequate catch records, and, in addition, an organization which provides some overview of the activities of all nations who wish to exploit living resources.

7. Provisions must ensure the opportunity to completely harvest the renewable resources at the appropriate level; that is to say, international standards must be applied in a manner that prevents the waste of renewable resources and provides for adequate opportunity for the full development and use of all fishery resources. A nation must not be allowed to prevent the use of fishery resources simply because these resources lie off its coast.

8. A combination of national and international enforcement of accepted rules must be provided for to prevent unfair treatment of any nation's

fishermen and to ensure that all fishermen operate according to accepted norms.

If the future brings us a broadly accepted set of rules governing both the catch of fish from the ocean and the operations of fishermen, then nations of the world will have the necessary guidelines to resolve disputes with little interference to their activities. In the absence of internationally accepted standards of conduct, one can expect unilateral actions by states extending jurisdiction over fisheries; continued dissent by those who refuse to accept such unilateral actions; increased interference in the development and use of the food resources of the sea; depletion of the living resources; and general chaos.

It seems obvious that the only sensible alternative is to accept principles along the lines suggested above and work toward a national world fishery regime that will give in the long term the maximum food from the sea and reasonable stability for fishermen and nations.

April 1971

REFERENCES

Alverson, D. L., and E. A. Schaefers. 1965. "Ocean Engineering: Its Application to the Harvest of Living Resources." *Ocean Science and Ocean Engineering, Transactions of the Joint Conference, Marine Technology Society and American Society of Limnology and Oceanography,* pp. 158–70.

Bogdanov, A. S. 1965. A Preliminary Summary Review of Some Data on Sea Productivity. Manuscript.

Chapman, W. M. 1965. "Food from the Ocean." *Proceedings of the Fourteenth Annual Meeting of the Agriculture Research Institute, National Academy of Sciences,* pp. 65–94.

———. 1966. "Ocean Fisheries: Status and Outlook." *Exploiting the Ocean, Transactions of the Second Annual Meeting of the Marine Technology Society,* pp. 15–27.

Finn, D. B. 1960. "Fish: The Great Potential Food Supply." *Fishing News International,* Vol. 1, No. 1, October.

Graham, H. W., and R. L. Edwards. 1962. "The World Biomass of Marine Fishes." In *Fish in Nutrition,* pp. 3–8. London: Fishing News, Ltd.

Hale, W. E., and D. F. Wittusen. 1971. *World Fisheries: A Tragedy of the Commons?* Woodrow Wilson Association Monograph Series in Public Affairs, No. 4. 63pp.

Johnson, James H. 1971. "Trends in World and Domestic Fisheries." *U.S. Naval Institute Proceedings* (June), pp. 25–35.

McKernan, D. L. 1968. "International Fishery Policy and the United States Fishing Industry." In DeWitt Gilbert (ed.), *The Future of the Fishing Industry of the United States,* pp. 248–58. University of Washington Publications in Fisheries, n.s. 4.

Meseck, G. 1962. "Importance of Fisheries Production and Utilization in the Food Economy." In *Fish in Nutrition,* pp. 23–27. London: Fishing News, Ltd.

Murphy, G. I. 1966. "Population Biology of the Pacific Sardine (*Sardinops caerulea*)," *Proceedings of the California Academy of Sciences,* 34:1–84.

Pike, S. T., and A. Spilhaus. 1965. *Marine Resources.* A report to the Committee on Natural Resources of the National Academy of Sciences–Natural Resources Council. NAS-NRC Publication 100-E: 1–8.

Riley, F. 1971. *Fisheries of the United States, 1970.* Current Fishery Statistics No. 5600. 79pp.

Schaefer, M. B. 1965. "The Potential Harvest of the Sea," *Transactions of the American Fisheries Society,* Vol. 94.

———, and D. C. Alverson. 1968. "World Fish Potentials." In DeWitt Gilbert (ed.), *The Future of the Fishing Industry of the United States,* pp. 81–85. University of Washington Publications in Fisheries, n.s. 4.

4

Some Thoughts on Fisheries and a New Conference on the Law of the Sea

WILLIAM T. BURKE

The subject I would like to examine here, albeit far more briefly than is appropriate, is the relationship between international fishery regulation and the next, presumably imminent, Law of the Sea Conference (LOS). The assumptions underlying this topic are that (1) a new conference on the law of the sea will probably be held in the foreseeable future, (2) such a conference will deal with some specific issues that bear on fisheries directly or indirectly, (3) the conference will seek to deal with many other issues, and (4) it is presently useful to speculate about the effects of such a meeting upon future fishery regulation.

Why bother to speculate about the relationship of such a conference to world fisheries? My assumption is that consideration of this subject now may assist by calling attention to some of the possible outcomes at such a meeting that could be harmful to the common interest in fisheries regulation. It is also perhaps helpful to begin now to envisage the changes in regulatory structure that might eventuate and to consider how to cope with these changes. Outcomes may be forestalled by calling attention to their excessive cost, and changes might be made to occur more smoothly or with less disruption by anticipation and planning. In any event some

William T. Burke is professor of law at the University of Washington. He has been actively involved in ocean law, serving as a consultant and member of many committees. Most recently he has been a member of the United States delegation to the Intergovernmental Oceanographic Commission; a member of the Fisheries Oceanography Advisory Committee, United States Department of State; and a member of the Ocean Affairs Board, National Academy of Sciences. He was formerly a lecturer at Yale Law School; professor of law, Ohio State University; and acting director, Mershon Social Science Program in National Security, Ohio State University.

For generous financial assistance to this and other studies pertaining to international fishery regulation, the author wishes to express his gratitude to Resources for the Future, Inc., Washington, D.C.

effort at foresight is necessary to realize any advantage to be gained from anticipation of change in regulatory structure, and change is about the only thing we are certain of these days.

For purposes of this inquiry I have tried to think and to organize the result in terms of certain phases in the process of decision employed in projecting international fishery regulation. The latter concept is here conceived broadly to include the unilateral actions of states, as in prescribing fishery limits or other boundaries affecting fisheries, as well as regulatory activity conducted in an organized, highly structured way by means of the international fishery commission or other instrument. Accordingly the categories used to depict phases of the decision process are in terms of who, for what goals, employing what assets, guided by what strategies, makes what decisions (achieves what outcomes) with what effects. "Who" refers to the identity of the decision-makers; "goals" refers to the objectives of fishery regulation; "assets" refers to the values employed in seeking goals; "strategies" calls attention to the means used in manipulating assets; "outcomes" denominates the immediate result of the decision process; "effects" is a reference to longer term consequences.

Decision-Makers

Clarification of Policy

Before discussing the future as it appears now, it is useful to focus upon desirable policy regarding participation in decision making at a forthcoming LOS Conference. More precisely, since this is a broad multilateral gathering composed of virtually all states on the globe, what kinds of decisions ought to be made by those participating?

Among the many issues to be negotiated there would seem to be least question, from a legal perspective, that boundary issues are most appropriately within the competence of a broad international conference. Decisions of fisheries limits or more broadly on territorial boundaries should be left to the uncertainties and time-consuming nature of the customary decision process, featuring unilateral claim and response as the prime technique. Such a policy preference not only accords with general expectations but also reflects the nature of fisheries. Limits and boundaries are not isolated or regional issues for the reason that fishing is now prosecuted by numerous states in the most distant reaches of the globe. Since the effects of boundary delimitation are so obviously inclusive in nature, it is appropriate that the permissibility of boundaries be determined inclusively and through an organized method.

For other controversies over fisheries, however, participation in decisions should vary in terms of the localized character of the basic problem. Decisions about distribution of the benefits, or the yield, of particular fisheries are necessarily affected by numerous factors unique to the particular situation. Decisions over conflicting claims in specific contexts should be left to the states most closely concerned therewith. At the same time, however, some factors important for decision are sufficiently common, even if they are not always applicable, that some general principles might be enunciated, and this task is appropriately performed through a multilateral conference. General principles regarding preferential rights might fall in the latter category.

Probable Trends in Decision

An initial question of particular interest is whether one or another significant aspect of fishery regulation around the globe will be promulgated by the states' participation in the next Law of the Sea Conference or whether some other method will be employed in addition to or instead of a multilateral conference. This query can be more narrowly put, as follows: Will states decide this and other questions by acting unilaterally or will they act in concert, by mutual agreement? The probable answer is that both methods will be employed. As a preliminary matter it is important to take note of the effect of the mere fact that LOS negotiations will be expected to occur. It seems to be the prevailing expectation that the LOS Conference will have among its very many agenda items those of the territorial sea and fishery limits. If it does become clear that the conference will address the latter question (as well as the former) this perception could influence the eventual substantive outcome. One contingency to consider is that a number of states, perhaps a large number, will anticipate the conference and prior to its convening will unilaterally extend their territorial sea or contiguous fishing zone or both. The purpose of such action would be, of course, to permit the tactic of arguing at the conference that no lesser limit for these regions is acceptable or feasible or, even, possible, in view of the pre-existing limit. The significance of these unilateral decisions might be, thus, to have some, perhaps important, effect on the range of choice for states at the LOS Conference. The nature of this effect would be to diminish the likelihood that certain limits, lesser than the limit or limits claimed by the several states, could attract the necessary number of votes at the conference. The end result could be to prejudice fatally any agreement at the conference on a particular limit.*

* In commenting on this, Dr. Francis Christy (personal communication) observes as follows: "A contrary effect can also be conceived—that states may *refrain* from unilateral claims until they can see whether or not (and how) they might benefit from alternatives presented at the conference." I agree that this effect may well occur. It also may be that states will use still another anticipatory strategy, namely of

An additional possible effect, though not too likely, is that the pattern of unilateral claims would be so repetitive among so many states that a new customary international law limit would be established. This seems too remote an eventuality to warrant more than mere mention at this stage.

In whatever manner states behave prior to the meeting, it is highly probable that the next LOS Conference will be charged with the burden of attempting to establish, by one or another method, a limit on exclusive coastal control over fisheries. The implications of this need some emphasis in terms of who would be making decisions. The most important one is that a very considerable number of states will participate in decision making that do not themselves have a coast, or, most likely, a single fishing vessel. This participation means that states which cannot themselves assert a claim to a fishing limit (though they can reject others' claims) are influential in determining the outcome of the community's organized decision process. This oddity aside, the import of this is that it seems difficult to determine how these states will vote for the numerous likely proposals at a conference. Since these states do not engage in fishing themselves, the factors they consider important for choice may vary a great deal from one state to another or even for the same state as the conference progresses. Another way of putting this is that for landlocked states the trade-offs may be especially arbitrary insofar as the specific issue is concerned. These states could turn out to be of pivotal importance on some crucial issues at the conference.

There is a pretty good chance (it seems more likely than not) that the conference will be unable to agree on a fishing limit. On such an assumption the alternatives for states might resemble those that presented themselves in 1958. As may be recalled, states were then unable to agree at a general LOS Conference on a fishing limit and it was agreed to try again at a conference focused on only issues pertinent thereto, that is, the territorial sea and a contiguous fishing zone. (The latter conference, held in Geneva in 1960, was also unsuccessful in reaching a decision on these issues.)

In sum it seems very likely that in the seventies, over one hundred states will convene at least twice to attempt to reach general agreement on fisheries limits.

I do not want to suggest by this emphasis on fishing limits that this is the only or the most important way of resolving the question of allocating fisheries around the world. However, a limit will be an important part of the solution, if one is reached, and it appears, in sum, that both unilateral claims and organized multilateral decisions will be employed

covert threats of unilaterally established wide limits to be carried out unless satisfactory benefits are obtained at the conference.

for decision making. In the end there is a good chance states can resolve this problem by general agreement but this is more likely to happen, in my opinion, at a conference dealing especially with this problem. And such a conference will probably not occur before the mid-seventies.

If states are unable to establish fishery limits at a large multilateral conference, because no single set of provisions can attract sufficient support, the ensuing situation could be most unhappy. The discouraging prospect would be even grimmer if this failure occurs at two successive conferences, the second devoted to this issue alone. The experience of the 1960s strongly suggests that states will extend their boundaries unilaterally, to increasing distances, and that there will be no recourse short of force for reversing this trend if, indeed, force could do so in any except particular instances. This development, in which states proceed unilaterally and outside organized processes, could be extremely dangerous if a large number of states deal with the fisheries questions only as an adjunct to territorial aggrandizement. In such a circumstance active violence cannot be ruled out as the arbiter of ensuing ocean disputes. Since such confrontations are usually resolved in favor of the stronger antagonist, there would appear to be advantage for many states in avoiding this possibility.

Objectives

In considering the future of fishery regulation around the world, the most important task is to specify the goals one prefers, that is, the objectives to be sought in attempting to resolve regulatory problems. Unless there is some conception of goal or objective, it obviously is very difficult to assess the various alternatives that might have more or less serious prospects of coming into existence.

Preservation of Minimum Order: The Avoidance of Violent Conflict

Except for sporadic and not overly consequential outbursts, fishery disputes have not produced significant violence between or among states since World War II. It is very much doubtful if at any time during this period nations in general seriously expected fishery disputes would be resolved by the application of force leading to serious loss of life for the contending sides. In some isolated situations this expectation of violence may well have prevailed but this does not change the more general attitude. Obviously force has been used, and often in support of claims that find extremely slight acceptance among states generally; however, it is still accurate to assert that few expect disputes to be resolved by superior force. If such were the case, it seems reasonably obvious that United States and Soviet relations with South American states would be different than they have been. And the United Kingdom's dispute with Iceland would have had a far different outcome.

The question is whether the future will be much different from the past in this regard. In other words, can we detect new conditions which might escalate these disputes to such levels of intensity that disputants or other nations would come to expect superior force to be the determining factor. Without elaborating, I doubt if the future will be much different, that is, I believe that expectations of violent outcomes will not prevail in this type of dispute. If this hypothesis turns out to be wrong, I suspect it will be because, as suggested above, fishery matters have become entangled with other issues more intimately connected with power considerations so that the outcome of a fishery dispute has such implications for power (i.e., military security) that the state with superior strength in terms of force will seek to make that strength determinative. This could occur if territorial limits become very exaggerated and supersede any notion of a separate fishing limit.

Accordingly one objective to be sought is that of preserving minimum order. In this specific context, of fishery regulation, the operational significance of this goal is to counsel against wide territorial limits which by their nature could be alleged to be dispositive of disputes over fisheries. It does not seem to me very likely that preserving minimum order will often be at stake in resolving other fishery disputes which present only such questions as distribution of yield or income from a fishery or provision for maintaining yields.

Wider Distribution of Benefits of Fishery Exploitation

It seems to me entirely possible that nations will begin to raise certain hitherto muted questions about the distribution of income from world fisheries. As everyone is now aware, the developing countries (LDC's) have mounted a furious and wholly serious campaign in the United Nations to seek a share in the benefits to be realized for mineral exploitation in the ocean beyond national jurisdiction. This goal finds eloquent expression in the concept that the sea bed (and resources thereof) beyond national jurisdiction are the "common heritage of mankind." The notion here is that all states should share in the income produced by exploitation of the natural resources of this area even if they do not themselves participate in the actual production. It may well be that some will suggest that this same notion should apply to fishery exploitation in the international common. (This notion has been dubbed "creeping common heritage," otherwise known as Francis Christy's Law.) The question for present purposes is how such a goal provides guidance in appraising outcomes for fishery regulation at a new LOS Conference.

There seem to be two rather obvious implications of this goal, one in connection with fishery limits and the other in regard to financing regulation beyond such limit.

With respect to the first of these, it may be that if high seas fisheries were also to be considered part of the common heritage of mankind, the effect might be to encourage a contraction of fishery limits, at least in those cases in which the coastal state had no strong interest in adjacent fisheries. It is difficult to generalize about effects otherwise. States which did not engage in full exploitation in their exclusive fishing zone might prefer to maintain a wide zone so that fishing rights therein might be disposed of for profit. Such states would also have to calculate, of course, that to keep this golden egg in untarnished condition there would probably need to be expenditures for management efforts. Since these costs might be substantial, especially if the state had an inadequate indigenous science base, some of these states might prefer to enlarge the area of common heritage and share therein. This would entail the contraction of national boundaries.

The other implication mentioned is that if fisheries beyond national jurisdiction were also to be denominated as the "common heritage of mankind," this would appear to place direct responsibility for the management on international institutions. This could presage a truly remarkable change in the arrangements for management. In such a circumstance, states engaging in exploitation might no longer feel required to expend resources on research and management, since presumably the common heritage should be regulated by the world community as such. This would differ vastly from the present situation in which the fishery commissions are almost completely the creatures of their creators and in nearly all cases without any significant independent power or initiative. If fish are part of the common heritage, then presumably their exploitation should provide some revenue for the states of the world. If these states wish, in turn, to maintain or increase this revenue flow then they will have to take measures to impose reasonable regulation upon the heritage in order to provide the conditions necessary for this goal. As you are aware, this process of regulation is very complex and difficult and could prove to be very costly.

This requirement of international action for direct management of fisheries is, very plainly, an extraordinary step to contemplate (although the step is frequently recommended) and there is every reason to doubt that states are at all inclined to take it. It is useful, however, to remind ourselves that a serious move toward realizing the goal of wider distribution of benefits from fishing, through the device of the "common heritage of mankind," entails drastic changes in international fisheries management.

The common heritage concept is, of course, only one means, and not an overly plausible one, of achieving the goal of improving distribution. Enlarging exclusive fishing zones is another method and will probably be more popular than any other at the LOS Conference.

Increased Production of Protein

In a world plagued by maldistribution of protein, it seems likely that enlarging the supply is a reasonable goal and that increasing the production of animal protein from the sea is desirable. The assumption is that if the total amount available is enlarged the chances are better that increased portions will go to those in need of it. This may not be true, of course, but the conditions determining consumption frequently have little to do with the ocean. It remains desirable policy therefore to seek this increase under circumstances that are favorable to distribution to protein-short areas.

This goal may be contrasted to that of decreasing the yield of animal protein from the sea. Measures having such effect do not seem in the common interest. Proposals for fishery regulation should, at least, be able to pass such a test of desirability. It is to be questioned, from this perspective, whether expanding exclusive fishery limits is acceptable community policy. To the extent that such expanded limits act as a deterrent to expansion of fishery efforts by developing states needing protein, or restrict continuing efforts, they contravene the common interest in increasing animal protein production from the sea.

Maintenance of Physical Yield from the Ocean

There is hardly anyone who is prepared to argue that a stock should be exploited to the point that it is unable to reproduce itself and maintain a fishery. Although argument has been made that this policy should in fact be implemented with respect to some species or stocks, it is not commonly regarded as a desirable general goal at least as an original proposition. Where costs of rehabilitating a stock exceed the benefits then, of course, there would be justification for destruction of a stock. Accordingly, with the latter exception, a minimum policy concerning physical yield is to avoid measures which permit this eventuality to occur.

There is more and more doubt attending the desirability of policies which are formulated in terms of maximum sustainable yield. It is more widely recognized now than ever before that a fishery regulated so as to permit a yield at this level may still be in very dire trouble and that far different regulation is required. Indeed the only real defense that can be made of this goal of management is that it may be a means of permitting still other goals to be achieved. It is more and more frequently recognized that MSY serves an important political purpose: indeed this purpose is perhaps its primary significance. As an independent management goal, therefore, the MSY leaves a great deal to be desired. What is required, instead, is focus upon the objectives which MSY is said to promote or to facilitate. It is not suggested that these objectives are indefensible, merely that MSY is meaningful primarily in terms of such objectives and as a quantity by itself is nearly meaningless.

Improved Economic Benefits from Fisheries

Probably not many would advocate that fishery regulation should aim primarily at enhancing the welfare of fish or at safeguarding the bureaucratic interests of government officials. The ultimate aim of fishery regulation is to improve the lot of people, and primarily (but not solely) of the people who endure the hazards of fishing or of investment in fish catching. On most occasions, but not all, the maximum contribution to this end is achieved by increasing the net yield which can be secured by catching and selling fish. This net yield itself is most likely to be enhanced by lowering the cost involved in catching the fish, but obviously other measures are relevant including those promoting use of unexploited species. The overall general interests of the community are promoted when resources are not unnecessarily devoted to fisheries which could be employed to meet other human needs.

This particular goal is becoming more and more significant on the international level, but it would surprise me if it were expressly sought at the next LOS Conference as a major objective of participants except in connection with fishing limits. With respect to management generally, it would be desirable if any international arrangements resulting from the conference did not pose a barrier to seeking this goal. But I suspect it is asking too much to expect that maximum economic yield will be enshrined explicitly as an international fishery management goal.

The key question at this stage is which of these objectives or combination thereof will be sought by the United States at a LOS Conference. On this point it is important to distinguish the overall government position from that advocated by individual components thereof. At the present time the objective sought by the United States appears to be a combination of protecting the economic interests of coastal states (the United States certainly being our prime concern) and deterring, hopefully, further extensions of coastal authority which would affect military interests. But there is reason to doubt that this combined position has any great promise of longevity. I believe the twin objectives will probably collapse rather quickly, to be replaced by the single idea of assuring United States military interests.

There is no doubt, I think, in the mind of most observers that the United States' LOS policy is most importantly affected by the Pentagon's view of military interests. This is the major reason for believing that of the two objectives mentioned, the military interest is paramount. In order to secure a narrow territorial sea or free transit through straits, the United States will very probably be willing to sacrifice its alleged fishery interests.

In fact there would not be a large element of sacrifice involved if the United States were willing to concur in a wide exclusive fishing zone (among other things) in return for a narrow territorial sea or free transit. As others have observed, notably Wib Chapman, the largest gainer from a two-hundred-mile fishing zone would probably be the United States. The reasoning is that this country already suffers from some of the disadvantages of a two-hundred-mile zone without enjoying any of its advantages; that is, the tuna industry as our main distant-water fleet already is confronted with a two-hundred-mile zone. In these circumstances, it would not seem overly difficult to accept a very wide fishery zone in return for transit rights, especially where the latter are prized the higher anyway. In these matters the strongest voice within the nation is military and the total United States position would probably defer to this voice rather than accept the more moderate views of those in the Department of State. Another way of stating this is that goals for fishery regulation very probably are irrelevant to United States policy objectives at an LOS Conference and will be sacrificed with equanimity if military interests might be served.

Assets

Is there anything consequential to be said regarding the values various states or groups thereof can utilize at a new LOS Conference? One obvious fact is that states will vary enormously in many of the vital components of power and other values. For example, in undifferentiated terms it is perfectly clear that the United States and the USSR will be the most powerful states at the meeting, even if power were measured only in elements pertinent to use and management of the marine environment. However, it is scarcely less obvious that in many instances at such a conference neither of these states will exert anywhere near the affirmative influence their general power position would suggest.

Thus it is reasonable speculation that the most potent asset or value a state can employ at the LOS Conference is its identification or participation with other states as part of a group sharing strong common interests on a particular issue or set of issues. As a general proposition neither the United States nor the USSR is a part of a very large group with such common interests—indeed, on numerous issues each of these two states whose interests are otherwise conflicting may find the other his only major ally. As is well known, however, the very numerous lesser developed countries (LDC's) do often find strong common interests composed of shared demands and expectations concerning their common plight. Accordingly, though in many respects any particular LDC state may be infinitesimal on the usual power scales, its actual influence at the LOS Conference will be substantial because of its alliance with numerous other LDC's.

It would not be surprising, partially because of the general strength of LDC's as such, if one or two or even several of these states came to have unusually marked influence, in terms of affecting outcomes, by reason of a leadership role among LDC's and a mediating role with developed states. Presumably, this would occur because of various unusual capacities of individual representatives which would make them pivotal figures in negotiations. However it is accounted for, some of the weakest states, in general power terms, may turn out to be the most powerful at the LOS Conference.

Disparity in control over wealth will be just as marked as with respect to power at the LOS Conference, but there is good reason to believe that this imbalance will be more important as a base of power than power itself. It does not seem likely that the United States will use wealth to affect the outcome for fisheries, but it is still interesting to note what appears to be happening or seems likely to happen.

The wealth I refer to is the near certain prospect that the United States (and probably a few other countries as well, including especially Japan) will be in the forefront in developing the natural resources of the sea bed beyond national jurisdiction. Unless treaty arrangements are concluded which alter the present situation, this development would almost certainly produce benefits solely for the producing enterprises and states except for the possible wider gains from lowering mineral prices. The reason the United States will play a dominant role in deep sea mineral development is primarily technological in nature but also partly because of high capital requirements and its importance as a consumer.

It is the offer to relinquish some of the benefits of this superior position that I refer to as a use of wealth. This effort to secure power by employing wealth is to be seen in President Nixon's statement of May 23, 1970, and the subsequent draft sea-bed treaty which was tabled as a position paper in August 1970, at the meeting of the United Nations Seabed Committee. In its present form the United States draft sea-bed treaty seeks to hold out the promise of a money pay-off to the LDC's as a means of achieving their agreement to a sea-bed regime which is thought to favor this nation's military interests and which does not seriously impinge on any private United States interests in the development of oil or minerals from the sea bed. This arrangement is implemented by limiting the continental shelf to a relatively narrow band adjacent to a coastal state and allocating a substantial part of the net revenues from the mineral resources of the sea-bed area beyond the shelf to an international authority to be spent for specified international purposes. Coastal states would retain the sole right to tax revenues out to the shelf limit, retaining a portion of no more than 50 percent and no less than 33⅓ percent of the net revenues in the area between the shelf limit and the continental margin (the so-called trusteeship zone), and all the net revenues in the areas beyond the margin would go to an international body to be spent for certain specified purposes. All activities beyond the shelf limit would be open to all states without restriction except for those activities specifically mentioned in the draft treaty (the only ones mentioned are exploration and exploitation of natural resources). The idea is to prohibit any controls over military uses of the ocean in the region beyond the shelf but at the same time to provide a pay-off for the coastal state and to individual states through an international organization to be created for this purpose. Incidentally it should be emphasized that my reference is to the *present* apparent United States position. Change is certainly not to be excluded. If the United States can secure its military interests by recognizing a *wide* continental shelf and no intermediate or trusteeship zone, this possibility becomes very likely.

One ingredient of this pay-off to the coastal state concerns fishery resources. The coastal state would retain complete disposition of the living resources (undefined in the draft) of the sea bed in the trusteeship area, that is, between the two-hundred-meter isobath and the continental margin. This is in contrast to minerals since the surplus income produced by these would be divided between the coastal state and the international authority.

Enlightenment and *skill* will, as implied above, sometimes prove to be important values for some states and on some issues these values would be extremely significant. Enlightenment is here used to refer to knowledge and, specifically, knowledge of the legal, political, and economic issues at the conference. As usual, enlightenment will be in rather short supply and some states will have a lot, and many states will have very little. In a sense this value is another measure of a state's involvement in the ocean since intense involvement will probably result in ample preparation for the LOS Conference. But I do not think a large supply of knowledge will be determinative on issues except under limited but important circumstances. Among the many LDC's at the conference, it is probable that the majority will not have anything near the expertise that a few of their number will possess. These few could well turn out to be leaders and their views and positions could carry unusual weight among their less informed allies. Because, as noted more below, the LDC's could be such a potent force, the positions advocated by their more informed leaders could be of decisive influence on some problems.

Strategies

Diplomatic Instrument

The principal strategies at the LOS Conference will be diplomatic in nature, comprising offers and counteroffers conveying suggested trades involving one preferred outcome for another. The trade-offs will also include, of course, promises concerning some matters which have no connection with the ocean. I do not intend to consider here the various parameters of national interests which might provide guidance on what trade-offs will be invoked by whom in fisheries matters. This task is such a complex job that it is best left with those who have resources to do it.

Discussion might be devoted to a number of elements of the diplomatic instrument as they relate to the LOS Conference, including: (1) size of delegation in relation to agenda size; (2) sequence of deliberations on agenda; (3) capabilities at determining influence factors; (4) knowledge and background on issues; (5) financing delegation and staff; (6) importance of bloc voting—Afro-Asian, Latin American, Soviet, Western, Soviet and Western, LDC's v. DC's; (7) relative number and weight of interests (trade-offs available).

For present purposes, however, discussion is limited to voting, which is easily the most important strategic problem at a conference. Under the usual ground rules at these conferences, important questions will require approval by a two thirds majority. On the assumption that all 126 member states of the United Nations participate on any given question, 84 votes would be required for adoption of an important provision and 43 votes would be required to defeat such adoption. With these quantities in mind, the significance of bloc voting becomes exceedingly obvious as does the nature of certain issues.

The most important probable split among states at the conference will not be on ideological lines, as was often the case at the earlier Geneva conferences, but on the familiar discrepancy between developed and developing states. Although the latter group is conventionally taken to include 77 members, it has 90 or more actual members, giving this bloc impressive weight. If the 77 vote together, they have enough votes to adopt any single provision. Even serious division within the group need not prevent a small majority of the 77 from blocking any proposal they deem unacceptable.

Since the 77 are such a dominant group, it is exceedingly important to envisage how they might operate. Here the most vital point is how issues are perceived at the conference. If an issue is perceived as an economic one primarily, as affecting the allocation of resources or the benefits of resources, it seems to me highly probable that the developing states will seek to act as a single group. It does not follow that they will always succeed in identifying a common interest which can be protected by shared action. Nonetheless, that an effort at group consensus is made will be extremely important even it if fails to achieve cohesiveness. For even if the group is divided, it may easily happen that one segment is sufficiently large to block any proposals thought unacceptable. Or one of the segments may be sufficiently large that it is conceivable that enough outgroup votes can be obtained to carry the day.

For states not part of the 77, including, primarily, the United States, the USSR, Canada, Eastern and Western Europe, Japan, and Australia, the most practical voting strategy is probably that of attempting to form a blocking third (plus one) at all feasible points. On very many issues, it seems clear that the dominant voting strength is in the LDC bloc. At the very least, or perhaps most, the effort needs to be made to prevent these states from imposing their will irrespective of common interests. The formation of a strong blocking third may open the way to reasonable compromise in addition to preventing provisions contrary to genuine common interests.

Economic Instrument

It was noted above that the United States is seeking to use its favorable wealth position as a means of obtaining a particular outcome at the next LOS meeting. Fishery resources are only a minor part of this insofar as the sea-bed treaty is concerned, since the main bargaining points in this regard have to do with oil and hard mineral resources. Moreover, the United States occupies an entirely different position with reference to exploitation of oil and minerals resources than it does with reference to fisheries. For the former the United States is by far the most intensive exploiter in the world and has interests throughout the world as a result of oil industry activities. This nation's fishing industry is, in contrast, and with a couple of notable exceptions, largely local in nature, seldom straying far from the United States coast. The United States has little actual control over world fisheries or any dominance therein.

It is nonetheless a provocative question, already mentioned briefly earlier, whether the United States will seek to employ the concept of coastal control over fisheries as another of the means of securing military interests. The parlay here would be to establish a zone for fisheries just as one has been proposed for minerals, that is, to assimilate fisheries *above* the continental shelf or other submarine zone to the minerals on the shelf below. In return for allocating both minerals and fisheries to the coastal state, the latter would agree to a narrow territorial sea of no more than twelve miles and permit free transit through it. In suggesting this possibility, which is not original, the thought is simply that United States officials may be willing to trade both fisheries and minerals for military advantage. There is very little evidence to suggest this tactic will work and much to suggest it will not.

This strategy could obviously have a number of variations. One might be to propose several fishing zones, such as an exclusive fishing zone beyond the territorial sea and beyond this an exclusive management zone. In the latter zone the coastal state would act as "trustee" for the world community in managing the fishery resources but would not have exclusive access.

It would not be wise to dismiss this idea too quickly. In 1958 the United States did propose an exclusive fishing zone as one means of retaining a narrow territorial sea. There is no reason to believe the latter is now regarded as any less important to the United States than it was in 1958.

One intriguing aspect of this possibility is that it is the exact reverse of the creeping jurisdiction notion that is so fondly held in the Pentagon. This notion is that extension of jurisdiction for one purpose inevitably leads to extension for another purpose. If the United States attempts to trade both minerals and fisheries for military security, it would mean that an extension of jurisdiction for several purposes was being used to limit jurisdiction for another. Such are the ironies of international diplomacy.

An economic strategy is relevant here for other reasons closely concerned with fisheries. If a modicum of rational calculation is conceivable in this context, it is not unlikely that some major bargaining considerations are economic in nature. In negotiating over fisheries zones, the major offers may well be examined in economic terms—how much cost is incurred if Deal *A* is accepted rather than *B*. Fishing limits and conditions of access to various zones will probably be appraised in terms of their effects on the cost of fishing. Coastal states who are seriously concerned with reaching agreement can bargain most effectively in terms of arrangements that seem likely to lessen the costs imposed by a wide limit. The wider the limit, the greater the importance attached to the costs of access to the area: how large a license fee; what stocks of fish will be open to foreign exploitation under license; when fishing will be permitted. Offers and counteroffers so expressed are instances of what I call economic instrument of policy.

Military Instrument

Although a major stake at the next LOS Conference is military power, it is not to be expected that military means will play any significant role at the conference itself. However, it is true that individual threats of coercion are not unknown in proceedings of this kind. Fisheries do not appear to be valued so highly as to occasion a threat of force, but it may be recalled that during the 1958 Geneva Law of the Sea Conference a threat of assassination was employed on at least one occasion. But it does not seem likely that any state would employ this instrument regularly (or at all in any situation where it was likely to become public knowledge).

Outcomes

What alternatives for fishery regulation could eventuate at a new LOS Conference? A review of these in relation to the goals we postulate for such regulation perhaps suggests some outcomes to avoid at such a conference.

It seems to me that two lines of development important for fishery regulation can be derived from the 1958 and 1960 Geneva conferences and that they could provide some guidance to likely happenings at a new conference. The first of these was the effort to reach agreement on a relatively narrow territorial sea by adopting the strategy of separating the issue of fisheries from the territorial sea issue and urging a wider limit for the former than for the latter. The aim in this strategy was to satisfy the military interest in safeguarding transit from coastal interference and at the same time indulge coastal states who felt a need for enlarging the extent of control over fisheries. These solutions put the emphasis on delimitation of zones in the ocean.

A second line of development begun at Geneva, and mostly left to lie dormant since, is expressed in the provision of Article 6 of the Convention on Fishing and Conservation of the Living Resources of the High Seas. This is the recognition of the "special interest" of the coastal state which, in this particular agreement, refers only to the "maintenance of the productivity of the living resources in any area of the high seas adjacent to its territorial sea." In terms of implementation of this convention, the coastal state's special interest has gone largely unrecognized. However, the various bilateral agreements negotiated among and between the United States, Japan, the Soviet Union, Poland, various European states, Indonesia, and Australia do seek in various ways to recognize a special interest of the coastal state. In part these arrangements are important in relation to the twelve-mile fishing limit, but they go much beyond to deal with operations within and beyond special fishing zones.

In mentioning these two features of recent international negotiations, the aim is to suggest that it is useful to view the forthcoming LOS negotiations as something of a contest between, or contrast of, these two different approaches. In a sense the suggestion of a modest exclusive fishing zone coupled with provisions for carefully defined preferential rights to stocks beyond such zone is the analogue to the two-hundred-meter continental shelf plus trusteeship zone which the United States is promoting for the sea bed and subsoil beyond the territorial sea. This idea of a narrow territorial sea or exclusive fishing zone plus preferential rights beyond is, in my opinion, to be contrasted sharply with proposals for a greatly enlarged territorial sea or fishing zone. These possibilities might be outlined as follows.

On the one side is the suggestion by the United States for a twelve-mile territorial sea, with issues concerning fisheries beyond resolved by means of recognition of certain preferential rights in the coastal state. (The United States twelve-mile territorial sea proposal is accompanied by a proposal for recognizing freedom of transit through and over straits.) In this view there would be no special zone for fishing rights but rather a set of formulas for disposing of particular stocks. It is the emphasis of this approach to deal flexibly with the adjustment of coastal interests, identifying these interests in a variety of ways and permitting some foreign fishing in the area adjacent to the territorial sea.

The opposing positions are various but, in the view here proposed, have in common a reliance on the notion of establishing specific limits for accommodating coastal and noncoastal rights. The limits can obviously be defined in a variety of plausible ways combining (or not) a territorial sea and an exclusive fishing zone in conjunction with preferential rights. Likely possibilities are a two-hundred-mile territorial sea, or a lesser territorial sea with an exclusive fishing zone out to two hundred miles. The extended zone could be measured in terms of the continental shelf, assuming this is somewhere defined as, for example, the sea bed out to two hundred meters or fifty miles, whichever is greater. It may be that some states would seek a territorial sea wider than twelve miles, but not two hundred, say thirty miles, plus a fishing zone for a further distance plus a system of certain detailed preferential rights out to two hundred miles. In this latter instance, the two contrasting approaches are merged although it might be said from the United States official view that its proposal was completely submerged. As noted earlier, it would not be a great surprise if the United States were willing to accede to such proposals providing free transit is recognized.

At this stage in time the prospects appear to be better for those taking the more simplistic view of using wide fishery limits alone without recourse to the greater complexities of preferential rights. A number of reasons might be cited for this former view. For one, a number of states already employ wide limits and some of these are bound to make a proposal for their general adoption at such a conference. There may be attractions in a proposal which both appears beautifully simple and at the same time gives the coastal state greater authority over a large area of ocean water. This latter feature would be extremely appealing to some of the developing states who might perceive the possibility of additional revenue from this source. Another reason is that states might be persuaded by the contention that there is ample precedent for a wide fishing zone. After all a great many states have exclusive fishing zones of twelve miles; hence, it is demonstrably true that *some* such zone is perfectly lawful. Moreover several states have had a very wide zone for many years and have not been successfully diverted from this path by the contravening objections. And to bolster this case, it might be emphasized by some that the method of international regulation of fisheries has not at all been a marked success, and no other plausible method appears to be in view.

It is not irrelevant to make explicit reiteration here that the United States may well place such great weight on satisfying its supposed military interests that it will be willing to accept a large exclusive fishing zone in exchange for arrangements assuring free transit in straits. This possibility adds credibility to speculation that a wide fishery limit may emerge as the most popular single choice at the next LOS Conference. However, in my opinion, the United States will not succeed in trading off a fisheries limit for freedom of transit. The LDC's will accept the former but not the latter.

Effects

An important task is to speculate about the potential long-term consequences of various *outcomes* at a LOS Conference insofar as fisheries are concerned. One such outcome may be that no agreement is reached on fishery issues.

If no agreement is reached on a limit for the territorial sea which is satisfactory for fishery purposes or on a fisheries limit, it seems probable that states will initially proceed to act unilaterally to promulgate a limit for fisheries purposes. Such a limit will probably constitute the boundary of an area within which the coastal state will dispose of complete control over all aspects of fishery exploitation, including especially the competence to determine who will get what portion of the permissible catch and under what conditions. The limit seems likely to be extensive, embracing all areas of even remote interest to the coastal state. It would certainly not be surprising if the limit were very commonly set at two hundred miles. While this limit would not take care of anadromous species, it would embrace many others except for their lateral movements into adjacent exclusive fishing zones.

The two-hundred-mile fishery limit could conceivably also be an outcome of an agreement at the LOS Conference or a subsequent meeting. In either case, whether it eventuates from negotiations or as a result of a failure to negotiate an acceptable limit, it is useful to inquire into the effect of such a limit upon the existing international fishery commissions. The following discussion considers first the effect of a two-hundred-mile limit and then, rather briefly, the effect of a modest limit coupled with preferential rights.

The impact of the two-hundred-mile limit on a fishery commission will vary, of course, depending on the proximity of the commission area and the stocks therein to coastal states, on the identity of parties to the commission, and probably on other pertinent factors. For some commissions, such as the whale and tuna, the stocks are caught throughout the ocean and coastal limits do not embrace anywhere near all the resource. Accordingly there is no feasible alternative to employing an entity inclusive enough to cover the entire stock being exploited. An entity meeting this description is one which is composed of all those states who engage in significant exploitation. No single coastal state could possibly serve this function.

The question for other fisheries is whether or not an expansion of the fisheries limit to two hundred miles would embrace a sufficient proportion of the exploited stock that effective coastal controls are possible. Apparently there are some very important fisheries which would be embraced by this seaward limit—the Georges Bank haddock, the entire fishery area of the Grand Banks, the fisheries off Norway, fisheries off South Africa, the Peruvian anchovy fishery, saury fisheries off the west coast of the United States, and practically all crab, lobster, and shrimp fisheries of the world. Important stocks would thus be subjected to the regulation of a single state during most of the harvestable stage.

The question is, then, what purpose would remain to be served by fishery commissions formerly having sole cognizance of any of these stocks. If the coastal state is fully competent, that is, legally authorized, to establish regulations limiting catch and effort (including even total exclusion if it wishes) then it would not seem that commissions any longer serve a purpose. The only reason for creating the commission in the first place was that no political authority existed which could adopt regulations extending to a sufficient proportion of the stock to be effective. International agreement was the only alternative.

It may be, however, that even a two-hundred-mile limit frequently does not include the whole of a stock in its exploitable range. This may occur because the stock normally occupies an area extending beyond the limit. Or it may be because the stock migrates into the area for a period but then moves on again to another region outside the two-hundred-mile limit. The movement may be into the waters of another two-hundred-mile limit state or it may be into the high seas or both. The situation of such fish would somewhat resemble that of anadromous fish which are subject to fishing both on the high seas and within the waters of one or more coastal states.

In this case it would appear that a commission, but not necessarily a pre-existing one, may continue to serve a purpose. There still will be situations in which coastal state regulation alone cannot be made effective and international regulation remains necessary. It may be, however, that there is no need to maintain a formal commission with a secretariat for this purpose, an annual meeting of interested states being sufficient to dispose of any of the problems. This latter pattern might be particularly applicable in those instances in which the stock moves only from one coastal state zone to another and back again. It could be under these circumstances that bilateral agreement, renewable annually or periodically, would be wholly adequate for the task without the need for any special intergovernmental mechanism. Certain situations, however, may require more formal arrangements because of the complexity of management needs—the International Pacific Salmon Fisheries Commission may illustrate this set of circumstances.

For those situations in which stocks move in and out of coastal zones, including high seas, something like the present arrangements (only much improved) may continue to be useful since the problem of regulation will strongly resemble that presently confronted only perhaps in lesser intensity. In these circumstances the fish are still vulnerable to uncontrolled access in the high seas as well as to controlled access within the fisheries zone. Unless the fish spent a sufficient period of time in the region of coastal states' control, this form of regulation would be inadequate. Accordingly it may still be useful to examine these present arrangements for the purpose of suggesting improvements in them for future application.

Another possibility requires mention. If states generally do extend their fishing limits drastically, as to two hundred miles, it could have a number of consequences prominent among which is that fishing states equipped to do distant-water fishing may increase their efforts to develop unexploited stocks which lie beyond the new limits. If this proceeds on a sufficient scale with a number of participating states, the result would very likely be a need for a regulatory program established by agreement among these states. It is not inconceivable, therefore, that extension of fishery limits may produce a need for new institutions in some parts of the world even as other institutions became unnecessary. It could be also that some existing mechanisms may feel pressure for improvement under these circumstances, for example, the Indian Ocean Fishery Commission.

Another consequence of enlarged fishery zones is that coastal states would seek to dispose of some or all of their right to exploit by means of selling such rights to others. That is, instead of actually attempting full exploitation of fisheries enclosed by the extended limit, the coastal state would prefer to realize revenue by permitting others to engage therein on payment of a fee of some sort. The conditions for permitting access, including most importantly the amount of consideration exchanged therefore, could conceivably be the occasions for negotiation between the coastal state and interested exploiting states or private exploiters. Although these arrangements could perhaps be most usefully concluded on a bilateral *ad hoc* basis, it could be that a wider institutional arrangement might prove more effective. It is even conceivable that outside fishing states seeking access to the fishery zones of an entire region might attempt to coordinate their activities and operations in order to promote efficiency and, perhaps, minimize bidding against each other for the rights in desirable places. The end result of this process could be, again, the creation of some institutional mechanism for safeguarding the various interests at stake. An organization of this type would probably differ very substantially from the fishery commissions we know now since the primary emphasis would be on accommodation of economic interests, leaving the scientific problem of safeguarding stocks to the coastal state involved. At the same time the coastal state would continue to have an interest in fishing methods and practices since data on this would be essential for its management system, hence coastal states might also be interested in a role in new institutions of this kind.

Accordingly the above discussion suggests that enlarged fishery zones might be accompanied by, or lead to, the continuation of some of the present international institutions for management, the creation of some new regional institutions composed of coastal states, and, possibly, also of distant-water states wishing to fish in the various state coastal zones, and the evolution of a need for new institutions for regulation of fishery exploitation beyond any area of coastal control.

But what if the assumptions above turn out to be incorrect and states succeed in reaching agreement on a moderate but satisfactory fishing limit coupled with measures for according a preference to coastal fishermen with regard to stocks beyond such limit? The question here is whether it is possible to anticipate enough detail in advance to speculate reasonably about this situation with respect to fishery commissions. Some comments still seem possible. One, the granting of preferential rights will probably be geographically determined, that is, the area of preferential right will be immediately outside the coastal states' boundary. In this circumstance it is entirely possible that an entire stock will be the object of several exclusive rights, since the stock moves between and among coastal areas, and it will accordingly continue to be necessary to enter into agreements to adjust competing claims and interests.

Second, the high seas pelagic fisheries would still present a problem and an international arrangement would still be at least as necessary as presently. As noted earlier, the necessity may extend to new situations not now anticipated.

A third possibility emerges from the potential situation in which the same species is subject to several exclusive rights, each in a different area, of course. It is possible to imagine that some states holding such rights will not themselves wish to engage in the fishery and instead seek to sell their right. States wishing to acquire such rights might deal with this situation in a number of ways. One is that they will not wish to bid against each other to acquire the preferred right but rather arrange among themselves to bid in such a way as to minimize the cost. It is conceivable that this practice could lead to some institutional means for disposing and acquiring fishery rights, with the coastal states also participating in the institution.

In sum the adoption of a moderate fishing limit, coupled with provision for preferential rights, may not be significantly different in gross outline than promulgation of very extensive fishery limits.

In one respect the preferential right approach may call for continued, sometimes complex, negotiations concerning allocation of rights to the coastal state vis-à-vis other states. If the coastal state does wish to exercise its right it may wish to do so only in part, or only with respect to particular stocks, and it is conceivable that an institutional means for dealing with this situation could be felt to be justified or might prove to be needed.

November 1970

5

Economic and Political Objectives in Fishery Management

JAMES A. CRUTCHFIELD

On the ground that no discussion of objectives is meaningful except in terms of their inherent social goals and of our actual capacity to achieve them, this paper is somewhat broader in coverage than its title suggests. The first section reviews the changes in attitude toward the objectives of fishery management over the past decade and indicates the need for further development in this critical area. The following section briefly evaluates recent extensions of the bioeconomic theory of fishing under open entry and under institutionally constrained conditions. The final section deals with changes in the political and administrative environment, with particular reference to modification of objectives and of management techniques in specific programs.

To avoid repetition, the term fishery management will be interpreted throughout to comprise regulation of fishing mortality, enhancement of natural fish production, and accelerated development of the knowledge and technology required to convert latent stocks into economically useful resources. Each carries with it, of course, an important research component.

Objectives: From Maximum Sustained Yield to Multiple Social Welfare Functions

The sixties produced a substantial literature in economics and political science dealing with the efficient and equitable allocation of resources

James A. Crutchfield is a professor of economics and public affairs at the University of Washington. His major area of teaching and research lies in the field of natural resource economics, with particular emphasis on marine resources. He was a member of the Presidential Commission on Marine Science, Engineering, and Resources and was chairman of the Panel on Marine Resources for the commission (1968–69). He has served extensively as consultant and technical assistant in field programs of fishery development for the Food and Agricultural Organization of the United Nations and the United Nations Development Program from 1957 through in the public sector of economies in which most productive activity is governed by private market forces. For reasons that need not be repeated here in detail, the fisheries present, in one form or another, all of the major causes of market-mechanism failure that call for public intervention. These include external costs imposed by one group of fishermen on another; aspects characteristic of public goods, such as research and development findings in which one person's use of the service does not diminish in any way its availability to others; and the potential for useful programs whose obvious benefits can be realized only by expenditures so large that no private enterprise would rationally undertake them—particularly if many of the benefits would accrue to others. There are, in short, good reasons for public management of fisheries in the broader sense defined above.

It is significant that only by joint effort of many disciplines have studies dealt successfully with definition of the public interest requiring governmental intervention with market forces and with the criteria for allocation of effort within the public sector. Primary producers of the literature have been economics and political science, with a strong admixture of sociology and philosophy. The combination reflects two important developments in the theory of public expenditure: recognition of the inevitability of distribution effects, often of such importance as to require modification of choices based solely on economic productivity; and recognition of strongly held social attitudes that may influence not only what is done but also how it is done and by what units of government.

One key prerequisite runs through this entire field: the necessity of shifting from single- to multiple-objective functions in evaluating public programs and policies, with full recognition of the fact that no single common denominator—physical or economic—permits simple comparison of alternatives. The following discussion of the objectives of fishery management is geared to this central theme.

It is to be hoped that the late 1960s marked the end of simplistic controversies over the relative merits of maximum sustained physical yield versus maximum net economic yield as the objectives of fishery management. The new generation of fishery scientists is well aware of the fact that achievement of a desired level of fishing mortality by deliberate proscription of efficient harvesting methods is wasteful, self-defeating, and devastating in its effects on technological progress. The enforcement difficulties and higher costs imposed by such programs are

---

1971. He is presently a member of the National Marine Fisheries Advisory Committee and of the National Academy of Science Committee on Aquatic Resources.

This paper was adapted from an address presented at the American Fisheries Society Centennial Celebration, New York City, September 14, 1970.

obvious. In addition, the history of fisheries regulated into forced inefficiency demonstrates that the resultant warping of the geographic and temporal pattern of activity usually reduces physical yield below that obtainable under a more rational management scheme (Crutchfield and Pontecorvo, 1969). Only the essential point need be reiterated. The economic success of a management program is measured not only by the output of useful fishery products but also by greater outputs of other goods that can be produced when labor and capital employed in the fishery are held to a practical minimum.

Resource economists, on the other hand, in progressing from theoretical models to practical analyses of operating fisheries, have discovered the need for a thorough understanding of (1) the biological underpinnings of the production functions and (2) the statistical characteristics of data relating yield to effort. A model designed to explain the behavior under exploitation of a slow-growing demersal stock might be utterly valueless as a guide to economic evaluation or public policy for an anadromous fishery. And it remains uncomfortably true in many fisheries that variances swallow up central tendencies to such an extent that both objectives and methods of management must be tailored to the case, economically as well as biologically.

Both fishery scientists and economists, as they have learned to communicate more effectively, have become increasingly aware of the need to develop multiple-objective functions. This task is not unique to fisheries management. The much more sophisticated literature on the socioeconomic aspects of water resource development and water quality management now tends heavily toward the definition of optimal systems involving multiple objectives incapable of expression in a common denominator. However destructive of simplicity they may be in modeling or in policy formation, these complications can no longer be swept under the table.

In a more formal terminology, we may denote the welfare "output" of a fishery management program as $U = U(Z_1, Z_2 \ldots Z_n)$ where the Z's are different desirable outcomes: for example, contribution to net economic output, income redistribution, balance of payments equalibrium, reduction in structural unemployment, freedom from arbitrary government action. Maximization of U when all aspects of utility are directly measurable in a common yardstick (dollars, for example) is difficult enough on empirical grounds. But when the Z's represent desired outcomes, each of which can be ordinally ranked *but not in terms of a single dimension*, then the essential marginal comparisons are even more precarious. We are left with a truly multidimensional set of public interest elements and, thus, with the necessity of making explicit choices among conflicting objectives.

A number of important attributes of this "social welfare function" approach to fisheries management deserve further emphasis. First, given the "barn door" variances in some of the critical parameters it is unlikely that a unique optimal solution can be found; thus a cluster of generally acceptable packages of goals and of programs for their achievement may well offer a choice to be determined ultimately by political palatability. Second, the political process may be improved substantially by locking up some objectives as constraints. Thus, improved economic efficiency in a heavily exploited fishery might be accepted as a primary objective—but subject to the requirement that existing levels of employment must not be reduced by more than *X* percent, or to a requirement that any reduction in fishing capacity be equiproportional for different types of gear or for different geographical areas. Third, the choice of goals and their implementation must include at least rough estimates of marginal information costs.

The most pervasive and troublesome of the nonefficiency objectives of public investment, including management of natural resources, is the redistribution of income. The need for public intervention usually (though not always) arises for reasons that preclude charging the costs of a program to the beneficiaries; therefore, the identification of losers and winners should be a part of the planning process, and some kind of income objective, whether it be to redistribute or otherwise, must be recognized.

Public investments in fisheries can alter income distribution among groups in at least four ways: by income level; by economic function; by geographic areas; or, intertemporally, by curtailing current output (sometimes over long periods) in order to provide greater catches later—for other people. Unfortunately, the social scientist can tell us little more than the fisheries scientist about what *should* happen. He can, however, supply two helpful collections of data. First, it may be possible to quantify the effects of alternative policies on income redistribution; second, the cost of achieving any kind or degree of income redistribution, in terms of real output foregone, can often be approximated.

No one familiar with the history of fishery management needs to be told that more policy is determined by the pressure of well-organized groups of winners than by the criteria of maintaining a sound condition of the stocks or of yielding greater net economic benefits to society. The vital needs are to recognize that costs and benefits fall on different shoulders, to identify the alternatives, and to make the final income objectives a matter of informed and open public decision rather than of pure happenstance or raw political pressure.

A second area of concern is the impact of a fishery program on employment, particularly in the local area concerned. Economists have insisted, quite properly, that the new jobs and incomes are to be regarded as costs from the national point of view if the labor and capital employed are drawn from other productive activities. This would always be the case, of course, under the assumption (usual in benefit-cost analysis) of full employment in the national economy.

Conventional procedures for estimating potential net economic yield (and, implicitly, of establishing an efficiency goal for management) thus assume that all factors of production excluded from the fishery, or forced to leave it, are employable elsewhere in the economy. Unfortunately, however, the economic, social, and political isolation of fishermen makes them particularly immobile, and the "opportunity cost" (i.e., other income foregone) of employing them in the fishery may approach zero. Consequently, both political expediency and the dictates of common humanity frequently require that employment in a fishery be larger than efficiency considerations alone would dictate, at least during some transition period. It may even be that efficiency criteria are satisfied if excessive labor inputs to the fishery are less costly than alternative ways of providing minimum acceptable living standards.

Economic and political considerations in fishery management assume greater importance when we move from the traditional analysis of a single species exploited with a single type of gear to situations in which interrelated species, different types of gear, and perhaps different nations are involved. No criterion based solely on maximization of physical yield makes sense in dealing with intermingled stocks and nonselective gear. If the two or more species are valued differently in the market, maximization of physical yield from the total biomass is clearly an inappropriate objective. If the situation is further complicated by external costs imposed by incompatible types of fishing gear, economic and political factors become even more critical. Moreover, the mobility of fishing units cannot be ignored in framing a management program for a single fishery. For example, a program that successfully limits fishing mortality by reducing excess capacity in the western North Atlantic cannot be appraised independently of the uses to which the displaced capital and labor are put. If the result of successful management in a single fishery is simply to shift excessive fishing effort to other areas and other species, the net social benefits, however defined, cannot be inferred from performance of the regulated operation alone.

Obviously, this line of reasoning leads to the uncomfortable conclusion that the framework for managing international fisheries must become progressively wider with every increase in the mobility and sweep efficiency of modern fishing equipment. And this interdependence by species and region must influence not only the techniques of management but the choice of objectives and the weighting of their inherent biological, economic, and political elements.

Another complicating factor in defining the objectives of international fishery management is the enormous range in per capita real incomes. From the standpoint of efficient biological conditioning and economic harvesting of the stocks, it might well be simpler and more desirable to limit the number of nations participating as well as the number of units and types of gear required to harvest at appropriate levels. Yet this will inevitably create situations where management must take over before contiguous underdeveloped nations have acquired the capital and technical skills required for participation. The trawl fishery off the west coast of Africa is a prime example. In such cases national and international market mechanisms may provide persistently inaccurate signals to industry; and even if the inadequacies of market price determination in an underdeveloped nation are ignored, the division of a restricted catch among fishermen from underdeveloped and developed nations raises a moral issue too important to be brushed away.

An equally intractable consideration in the management of international fisheries is the problem of disparate social-value systems. Even in the simplest case where the activities of fishing industries are carried on in well-organized private markets, differences in tastes and preferences for end products, coupled with varying relative prices for inputs, would give rise to different optimal levels of fishing intensity and of gear mix in each participating country. Where national economies are organized on radically different principles, as in the case of the socialist nations, comparisons of alternative arrangements for fishery management in monetary terms are virtually meaningless. A bargaining range may yet exist within which all participating nations will be better off under regulation than under unrestricted fishing, but there is no single economic or social dimension common to all parties in which alternatives can be compared.

The objectives of a country participating in a marine fishery may be materially influenced by its current balance of payments position. Economists do not contest seriously the global advantages of a political environment in which the full advantages of international specialization and relatively free trade might be realized. But in the real world, full of frictions and obstacles to the movement of both goods and services, nations heavily dependent on the export of fishery products may place greater emphasis on the gross value of catches than on the economic benefits derived. It may even be perfectly rational, from a national standpoint, for Nation *A* to press for a larger share of a much diminished total physical and gross economic yield for all participants (but only if the other participants are content to let *A* play its own game).

In short, the objectives of international fishery management must be modified to accommodate different national objectives which in turn are based on different systems of social and economic organization, different alternatives for dealing with structural unemployment, and different degrees of dependence on fishery exports to meet current foreign exchange requirements. Since international trade differs in degree rather than in kind from interregional trade, it should come as no surprise that states and regional coalitions within a single nation advance similar arguments for modification of any management goals couched solely in terms of net economic benefits to the nation.

Moving even further afield, the goals and methods of fishery management may be constrained by political values associated with "the way things are done." The present fragmentation of authority over fisheries among state and federal agencies has had a demoralizing effect on the efficiency of both industry and public management. Yet, in framing remedies for an obviously undesirable situation, it must be accepted that a great many people might prefer to retain state control (while presumably hoping for greatly improved coordination) even if they were convinced that consolidation of all marine fishery management authority in federal hands would be more efficient.

Obviously, the key to a successful public program in fisheries lies in the ability to weigh these various objectives in some way that serves the public interest. Having decided what public is to be served, how can we choose rationally among alternative combinations? The literature on this critical issue is voluminous and remarkably untidy.*

One view simply takes economic efficiency as the dominant criterion, to be modified by secondary objectives only after explicit recognition of their economic costs (e.g., Mishan, 1967). A more sophisticated version would qualify this unequivocal acceptance of given essential weights by inferring them from past choices. Thus, marginal income tax rates might be assumed to indicate the importance attached by society to redistribution of income (Eckstein, 1961; Haveman, 1965). Others choose to regard the weighting system as an outcome of the political bargaining process. An intriguing outgrowth of this approach is the development of models of government decision making (including

* For an excellent review of the literature and the issues, see *The Analysis and Evaluation of Public Expenditures: The PPB System,* Compendium of Papers Submitted to the Subcommittee on Economy in Government of the Joint Economic Committee, Congress of the United States. Washington, D.C.: Government Printing Office, 1969. Vols. 1–3.

choice of objectives) considered as analogous to the functioning of the market mechanism in the economic sphere (e.g., Maass et al., 1961; McKean, 1968). A further, and more convincing, elaboration narrows the actors on the scene to socioeconomic pressure blocs, with the need for coalition as the effective (and efficient) constraint on government action.

It is clear that social scientists are far from consensus over the mechanisms of choice in public management and investment. But perhaps common sense may yet prevail where theory thus far has failed. There are answers to the question, "What is the public interest?" There are measurable magnitudes that narrow the range of choice. And there are ways of making intelligent and responsible decisions among the noncomparable alternatives that remain. Identification of multiple goals is the essential *first* step toward efficient formulation of fishery programs.

The efficient use of any natural resource cannot be denied as the prime objective of public policy, and it remains true that economically efficient management provides greater elbowroom to finance secondary objectives. But the growing conviction among social scientists is that valuation of a natural resource cannot be compressed into a single dimension without, in one economist's expressive phrase, "submerging real issues behind a façade of faulty measurement" (Steiner, 1969).

The Theoretical Basis for Management

As indicated above, the objectives to be expected of a modern fishery management program are bounded by the means of achieving them. To clarify that issue, this section deals very briefly with recent developments in the economic theory of fisheries and their significance in defining goals for public policy.

Social scientists acknowledge freely that the major difficulties in developing adequate models for exploited fisheries lie in the field of population dynamics. In the economist's terminology, the construction of production functions is critically dependent upon the analytical validity and accuracy of the biometric model. In addition, costs of production are heavily dependent on the ratio of fishing time to time at sea; efficient deployment of harvesting capacity requires data for temporal and areal distribution in addition to a gross yield estimate. Given these figures, the valuation of alternatives emerging from a properly specified biological model of an exploited fishery is no more difficult than the valuation of future benefit streams and of capital and operating costs associated with most other kinds of public investment.

A number of recent articles have clarified and improved the basic long-run bioeconomic models which initiated the dialogue among economists and fishery scientists (e.g., Gordon, 1954; Crutchfield and Zellner, 1961; and Christy and Scott, 1966). In particular, Turvey (1964), Smith (1969), Southey (1969), and Paulaha (1970) have clarified the nature of the socially optimal rate and techniques of fishing, given the conventional assumptions of a negatively sloped demand function for fish and a positively sloped supply function for inputs to a particular fishery. Under these conditions, maximization of economic rent from the resource alone will not produce the desired result. The optimal economic solution will devolve from that rate and type of fishing effort which maximizes the sum of consumers' surplus, economic rent, and producers' surplus.

These writers (and others) have also pointed out that optimal management strategies require separation of externalities arising out of mesh size, the impact of the activities of one fishing unit on the catch rates of others via stock effects, and congestion on favored fishing areas. (The last problem includes, of course, the fairly frequent case of incompatible types of fishing gear.) All of these modifications of the basic long-run model must enter into any management program that includes, explicitly or implicitly, improved economic performance as one element of the objective function. The writers mentioned have additionally clarified the discussion by recasting the models in a more conventional microeconomic theory of the firm.

The bioeconomic models have been even more significantly elaborated by extension of the analysis to mechanisms for temporal adjustment. As all previous treatments have emphasized, equilibrium in a commercial fishery requires simultaneous satisfaction of two sets of conditions: one relates to the size and composition of the biomass and, therefore, to its rate of change; the other relates to costs and monetary returns associated with the number and technical structure of the exploiting units. Analysis of the way in which a commercial fishery moves from one equilibrium position to another (and the possibility that some equilibrium positions may be unstable) requires analysis of the differential response mechanisms both of the stock and of the industry operating on it. The economists cited above, the staff of the Division of Economic Analysis of the National Marine Fisheries Service (formerly Bureau of Commercial Fisheries), and a host of population dynamicists in fisheries science have contributed to progress in this area.

Unfortunately, we are a long way from resolving the issues raised by the shift in emphasis from steady-state models to those involving transition periods, particularly as measured by operational usefulness in development and management of real fisheries. The greatest difficulties are likely to arise in developing satisfactory short period biological models of dynamic adjustment, particularly where the fish stocks in question are subject to a wide range of hazards additional to the harvesting by man. The magnitude of the variances in the analysis of exploited fishery populations is not an encouraging omen for the prompt development of such models.

This is particularly unfortunate, since only when considerations of steady-state equilibrium give way to the dynamics of transition periods can economics make one of its most effective contributions toward the formulation of goals. In brief, investment in fish stocks, as in any other assets, demands a reduction in the current output of economic goods to achieve greater output in the future. But since the public definitely prefers consumption now rather than later, and since other types of capital can be used now to expand output, future benefits must be discounted at some appropriate rate to determine whether, and at what point in time, fishery investments are worth their cost. This would present no problem if the time profile of benefits and costs from fishery management were identical, or nearly so; but since benefits typically lag costs by a substantial margin, the discounting to permit comparison is critical to economic appraisal of alternative management schemes.

In practice, rehabilitation of a depleted population inevitably involves some short-run sacrifice by the industry. The proper rate of rehabilitation is essentially an economic question, but most of the necessary data must come from the short-run response patterns of a biological model. If fishery management is defined to include enhancement and accelerated development of latent stocks as well as regulation of overfishing, it is again necessary to convert benefits and costs to a basis of present value in judging the economic feasibility of individual programs and in making a proper selection among alternative programs under budget constraints.

At least one group of fishery economists has argued that these short term problems are so intractable that the traditional economic insistence on maximum economic yield rather than maximum sustained physical yield is meaningless in framing practical political and administrative actions. In general, as the pressure on a heavily exploited species increases, a less significant economic cost will attach to the acceptance of efficient harvesting at maximum sustained yield rather than maximization of net economic benefits as a primary target. Since we can define accurately the stock conditions required for maximum sustained yield for relatively few fisheries, whatever difference there may be is likely to be swamped by the inadequacy of the data. Other economists are not prepared to relinquish their theoretical position so readily; but as a pragmatic matter there is something to be said for accepting the modified goal as long as the decision-making hierarchy still thinks largely in terms of maximization of physical output of fish.

From the standpoint of potential management capability, the usefulness of these refinements in the bioeconomic theory of fishing is extended tremendously by developments in techniques of data collection and manipulation. The economics of information are too often neglected in the evaluation of scientific programs, and fisheries management is no exception. The crucial issue for management is not how many statistics can be used by fisheries science but, rather, how few are essential to harvest most of the benefits of *timely* action. Recent developments in the use of simulation techniques to approximate the behavior of fish stocks and of segments of the fishing industry offer great promise in identifying these critical information requirements. By permitting us to test an almost endless variety of alternative formulations at low cost, they may enable us to bypass some formidable obstacles.

The computer obviously cannot make decisions, nor can it convert bad data into good; economic uncertainty cannot improve biological ignorance. But in studying the interaction of economic motivation and the productive capability of living resources, the ability to construct models of reasonable predictive value on the basis of relatively limited data would appear to offer an exciting new range of options in fishery management. Perhaps more important, it may represent the only way in which fishery science can hope to counter the appalling speed with which the misuse of marine fisheries can develop. Today's vessels simply have too much speed, flexibility, and range to allow the scientist the leisure to develop, slowly and systematically, the information systems which traditionally have preceded the formulation of fishery policy.

The urgent need is for reasonably accurate first estimates of the yield potential from latent fisheries before the avalanche hits. If the painful and all-too-familiar process of overinvestment, severe economic loss, and retrenchment is to be avoided, some method must be found to obtain at least a reasonable investment target for the exploiting industry. Recent work (e.g., Alverson, 1969) suggests that both the requisite analytical basis (and, in the very near future, the technology) may be at hand.

Development in Public Administration

Faint but unmistakable brightening on both the national and the international scene portends the dawning of a new era in the scope of fisheries programs. Obviously, the web of conflict and contradiction that now characterizes the administration of fishery management in the United States will not be untangled overnight—nor even over a generation. Nevertheless, the signs are clear that decision-makers in industry and government are becoming aware of the controversies over physical yield versus economic benefits as primary objectives of management. Also, a new breed of fishery specialists, thoroughly familiar with the literature behind the controversy, seems much better prepared to deal with the realities of program formulation under political constraints. It may be that even greater relative gains in realism and sophistication have been achieved in the theoretical and applied work of fisheries economists.

The results are evident in specific and fairly important instances. For example, the Department of Fisheries of Canada has undertaken a program of economic rationalization of the British Columbia salmon fishery. It would be hard to find a fishery in which economic inefficiency, unemployment, and the politics of gear conflicts add more complications to an already difficult technical management problem. This offers a particularly challenging test of the economists' position that major economic gains can be achieved by grafting additional goals and regulatory techniques onto reasonably successful biological programs. The stakes are substantial indeed, since the salmon fisheries of the eastern North Pacific are among the most valuable in the world. On a stage of such magnitude, a dramatic demonstration of the validity of objectives more realistic than maximum physical yield might well initiate a major revision of government attitudes. It is much too early to assess the impact of the British Columbia program, but it has sufficient support by government and industry to prove a real test of the multiple-objective concept.

More long term importance attaches to a study of alternative regulatory regimes undertaken by a standing committee of the International Commission for the Northwest Atlantic Fisheries (the outgrowth of studies by a working group of fishery biologists and economists from member nations of ICNAF and the Northeast Atlantic Fisheries Commission). Their report to ICNAF stressed the growing danger of overfishing of the North Atlantic demersals in both biological and economic terms, but with particular emphasis on the potential economic threat to industries involved and on the need for a regulatory framework encompassing a wide range of national attitudes and problems. The wheels of multinational fishery commissions grind even more slowly than those of the gods, and no formal recommendations have emerged which would indicate how far ICNAF is prepared to go in implementing the national-quota program recommended by the working group, or some variant of it. Nevertheless, it is truly heartening that this prestigious body would even consider such broad-spectrum regulation, since it is faced with every complexity to be expected in a fishery shared by fourteen nations with widely different economic, political, and social structures.

American policy for fishery management is beginning to respond to a number of quite separate forces, each adding pressure for a reformulation of concepts and methods of management. The report of the Commission on Marine Science, Engineering, and Resources (1969) called for a reorientation of both regulatory and development efforts, with emphasis on the long-run economic viability of the fishing industry. The commission also stressed (as does every knowledgeable observer of the American fisheries scene) that no real progress can be made in refining objectives or in carrying them out as long as authority remains divided among federal and state agencies. The chaotic state of fisheries statistics and the paralysis of unfocused authority preclude formulation of management activities on a uniform basis.

The commission recommended a solution which has proved effective in dealing with water and air pollution: that is, legislation under which the states retain paramount authority for the formulation and implementation of regulatory programs subject to federal jurisdiction if the states, individually or in concert, abdicate their responsibility. The concept would be more difficult to apply to the fisheries, since so many of the more important cases calling for intervention by government involve more than one state. Nevertheless, federal pressure to handle such problems through interstate compacts with real muscle could substantially increase the effectiveness with which logically defined fisheries stocks are exploited and managed.

The fisheries will also be affected, significantly though obliquely, by new legal developments in the coastal zone. It is manifestly impossible to maintain the arrangement of state jurisdiction out to three miles superseded by federal control from three to twelve miles. Some kind of rapprochement is required; and that very need presents a remarkable opportunity to develop more uniform objectives and a more effective state-federal partnership than exist at present.

The growing interest of both Congress and the Executive Branch in management of coastal and estuarine areas seems certain to produce major changes in administrative arrangements to meet pressingly important social and economic needs. Whatever the form of the new coastal-zone authorities that may emerge, effective management of the inshore environment must be integrated with management of the fishery resources in these areas. The fisheries are (and will probably remain) the most important single economic activity adversely effected by pollution, careless land use, and the destruction of natural habitat. But the fishery position must be evaluated within the framework of a more inclusive management unit responsible for all activities within the nearshore and estuarine environment. In terms of both net economic output and social outputs not measurable in economic terms, fisheries must be able to compete with increasingly insistent demands on the land-water complex in these critical zones.

Running through all of these developments are a number of common themes and unanswered questions. Both the Bureau of the Budget and the Congress are more and more insistently demanding identification of economic and social outputs in justification of fisheries programs. It is clear that state programs, particularly those dependent upon federal financial and technical support, must eventually meet the same criteria. Increasingly severe criticism is being leveled at the economic inefficiency of both developmental and regulatory activities; at subsidy programs that do not lead to a self-supporting, viable industry; and at misuse of the fisheries as a device to postpone long-run solutions to structural unemployment problems in coastal regions. Unfortunately, the more visible federal fishery agency has borne the brunt of the dissatisfaction, although in most cases the jurisdiction in question rests with the states alone under present legislation and practice.

To some extent, such criticisms of fishery programs are part of a broader feeling of frustration with government support of science in general—an attitude not entirely without foundation but one which might in the extreme have serious consequence to the nation as a whole. In fisheries, for example, it is analytically incorrect and potentially highly disruptive to insist upon application of the same basic economic and political evaluation techniques to mission-oriented development programs as to more fundamental research. For the former, it is usually possible to identify the probabilities of success, the economic benefits to be generated, and the recipients of those benefits. In such instances, benefit-cost analysis (or some other members of its family of economic and quasi-economic evaluation procedures) is directly applicable and useful. For discipline-oriented scientific activity, however, and for the development of fundamental technology applicable to a wide range of fishery activities, it is simply impossible, in theory or practice, to identify either the total magnitude or the timing of ultimate benefits. It is even less likely that the beneficiaries can be identified by class, much less in total.

Yet these are the very areas in which the strongest case can be made for governmental participation in research and development; and the breadth and quality of such programs are critically important, both to the establishment of objectives and to the precision with which they are implemented in the management of fishery resources. Rigid application of inappropriate economic criteria can easily twist scientific effort in the direction of short term, mission-oriented projects and programs for which clear-cut results "satisfying to the Bureau of the Budget" can be made available year by year. Over time, this short-run policy could dry up to an alarming degree the flow of cost-reducing basic information which must undergird the wise management of heavily exploited fisheries and the efficient harvesting of presently underutilized stocks.

Space permits only brief mention of one of the most formidable obstacles to the incorporation of economic, social, and political elements into a balanced set of objectives for fishery management: the lack of capability in these areas within the responsible agencies.

The National Marine Fisheries Service, after years of neglect of such matters, has developed excellent in-house economic capability, and its Division of Economic Research is beginning to generate the kinds of basic information required to set quantitative economic goals and criteria for both regulatory and developmental activities of the Service.

Unfortunately, nothing remotely comparable is to be found in any of the state governments. The Pacific coast states, including Alaska, produce a substantial proportion of the total value of American-flag fisheries; yet none of the state fishery agencies has a staff economist, nor is there provision for regularly obtaining the services of economists from other branches of state government. All have enjoyed the benefits of occasional *ad hoc* consultations with social scientists in universities, but these are no substitute for continuous involvement with the agency, the industry, and the broad field of fisheries science.

The international agencies are only slightly better off. The Food and Agriculture Organization of the United Nations and the United Nations Development Program, striving to accelerate the pace of growth in the underdeveloped nations of the world, could hardly avoid economic and political objectives in the formulation of fishery projects. FAO has attached a small but highly professional staff of economists to its fisheries department, and can also draw on the considerable social science capability of the organization's other departments. On the other hand, the international commissions charged with administering the existing programs of fishery management on the high seas have no social scientists on their own staffs, though they can draw on the services of specialists from the fishery agencies of the participating nations.

In short, one can hardly criticize fishery administrations at state, national, and international levels for failure to develop properly weighted multiple-objective programs when they have had no specialists competent to deal with such issues. Indeed, comparable problems are far from solution even in fields where economic and political analyses have long been an integral part of management. The vital importance of these elements must be recognized and matched by adequate staffing of the planning and executive agencies if a balanced program of fishery regulation, enhancement, and development is ever to be achieved.

October 1970

REFERENCES

Alverson, D. L. 1969. "Demersal Fish Explorations in the Northeastern Pacific Ocean: An Evaluation of Exploratory Fishing Methods and Analytical Approaches to Stock Size and Yield Forecast." *Journal of the Fisheries Research Board of Canada,* 26: 1985–2001.

Christy, F., and A. D. Scott. 1965. *The Common Wealth in Ocean Fisheries.* Baltimore, Md.: The Johns Hopkins Press.

Commission on Marine Science, Engineering, and Resources. 1969. *Our Nation and the Sea,* Report of the COMSER. Washington, D.C.: Government Printing Office.

Crutchfield, J. A., and G. Pontecorvo. 1969. *Pacific Salmon Fisheries: A Study of Irrational Conservation.* Baltimore, Md.: The Johns Hopkins Press for Resources for the Future, Inc.

Crutchfield, J. A., and A. Zellner. 1961. *Economic Aspects of the Pacific Halibut Fishery.* Fishery Industrial Research, Vol. 1. Washington, D.C.: U.S. Department of the Interior.

Eckstein, O. 1961. "A Survey of Public Investment Criteria." In J. M. Buchanan (ed.), *Public Finances: Needs, Sources, and Utilization.* Princeton, N.J.: Princeton University Press for National Bureau of Economic Research.

Gordon, H. 1954. "The Economic Theory of a Common Property Resource." *Journal of Political Economy,* 62 (April): 124–42.

Haveman, R. H. 1965. *Water Resource Investment and the Public Interest.* Nashville, Tenn.: Vanderbilt University Press.

Maass, A., et al. 1962. *Design of Water Resource Systems,* pp. 565–604 *passim.* Cambridge, Mass.: Harvard University Press.

McKean, R. N. 1968. *Public Spending.* New York: McGraw-Hill Book Company.

Mishan, E. J. 1967. "Criteria for Public Investment: Some Simplifying Assumptions." *Journal of Political Economy,* 75 (April): 139–46.

Paulaha, D. F. 1970. "A General Economic Model for Commercial Fisheries and Its Application to the California Anchovy Fishery." Ph.D. dissertation, University of Washington, Seattle.

Smith, V. L. 1969. "On Models of Commercial Fishing." *Journal of Political Economy,* 77 (Pt. 1): 181–98.

Southey, O. 1969. "Studies in Fisheries Economics." Ph.D. dissertation, University of British Columbia, Vancouver.

Steiner, P. O. 1969. "The Public Sector and the Public Interest." Pp. 13–45 in *The Analysis and Evaluation of Public Expenditures: The PPB System,* Compendium of Papers Submitted to the Subcommittee on Economy in Government of the Joint Economic Committee, Congress of the United States. Washington, D.C.: Government Printing Office.

Turvey, R. 1964. "Optimization in Fishery Regulation." *American Economic Review,* 54 (March): 64–76.

6

International Arrangements for the Management of Tuna: A World Resource

JAMES JOSEPH

The fact that a large share of the present world's population suffers from a deficiency of animal protein in their diet and that world population is continuing to increase in an explosive manner creates a tremendous need for the development of new sources of animal protein. The oceans of the world have been looked to as a great storehouse of animal protein and have frequently been suggested as a panacea to the problem of feeding a growing, human biomass. However, estimates of the sea's potential to do this have differed greatly (Schaefer, 1968; Chapman, 1970c; Ryther, 1969). Some indicate that little more than the current level of protein production from the sea can be expected; others, more optimistic, indicate production can be increased twenty-fold over current levels. The true potential production from the sea is unknown. It is certain, however, that if the sea is to supply a major share of the protein needs of a burgeoning population the catch of fish must be increased and the production of currently exploited populations must be maximized and sustained.

The production of fish from the oceans, inland seas, and fresh waters of the world has increased on the average 7 percent per year for the last twenty years. In 1950 the world catch of fish and shellfish was approximately twenty million metric tons and by 1969 was near seventy million metric tons. Over this same period no other basic food commodity has increased at even approximately as great a rate. During the same period of time the rate of increase of the world population has been approxi-

James Joseph is the director of the Inter-American Tropical Tuna Commission at La Jolla, California. He has been employed in several positions with the tuna commission, is chairman of the Food and Agriculture Organization's Expert Panel for the Facilitation of Tuna Research, and has responsibility for providing scientific advice on stocks of tuna in the eastern tropical Pacific.

mately 2 percent per year. If the world catch of fish continues to increase as it has in recent years, it has been estimated that by 1980 the sea will be able to provide the necessary animal protein requirements for the entire world population at that time (Schaefer, 1968). If the catch of fish does not continue to increase, then the animal protein gap will widen.

To attain the full potential from the seas' animal resources will require two things: first, an evaluation of the latent resources of the sea and the development of methods to harvest them; second, an understanding of the effect of man's intervention on the presently exploited stocks of fish and the implementation of management measures based upon this understanding.

This paper will deal with the latter subject but will be concerned with high seas tuna resources, which are of a truly international nature. I shall attempt to review briefly how much tuna is taken annually, by which countries, and from which oceans. A brief review of the current condition of the tuna stocks will be given along with some future prospects for exploitation. Present international arrangements for the scientific study and subsequent management of these resources will be discussed, followed by comments on how these present arrangements are succeeding to date and how they can be modified to better accomplish their purpose in the future.

To suppose that tuna is fished with the idea of contributing protein directly to a protein deficient sector of the population is a spurious contention at the least. Tuna is a high-priced commodity and is caught for its commercial value. The return from this catch of course will provide capital with which essential items including protein can be purchased. The present paper is written with this understanding.

The World Fishery for Tuna

The Fish

Though there has been and still is a great deal of disagreement on what tuna are, they are generally considered to be members of the family Scombridae which includes a large number of species, some of which are of commercial importance and some of which are not. This paper will deal primarily with the five * most important species of tuna: yellowfin (*Thunnus albacares*), skipjack (*Katsuwonus pelamis*), albacore (*Thunnus alalunga*), bigeye (*Thunnus obesus*), and bluefin (*Thunnus* spp.). These species comprise approximately 90 percent of the total world catch of all tuna-like fishes.

* Six, if the bluefin is considered as two separate species—northern (*Thunnus thynnus*), and southern (*Thunnus maccoyii*)—which they are taxonomically.

Tuna occur throughout all of the major temperate and tropical oceans of the world, primarily between 35°N and 30°S. They appear to congregate near the surface waters close to coastlines of islands and major land masses, and on the high seas near current interfaces and other areas where the upper mixed layer is shallow. In areas other than this they appear to be distributed more deeply throughout a thicker upper mixed layer.

With reference to their geographical distribution the five major species are generally categorized as tropical or temperate. As the name implies the tropical species, yellowfin, skipjack, and bigeye, occur in warmer waters than the others. Their distribution appears to be continuous, or nearly so, through the Pacific, Atlantic, and Indian oceans. Biologically these animals seem to be rather similar. Their metabolic rates are rather high, they grow rapidly, swim rapidly, and seem to spawn over vast geographic areas. The yellowfin and bigeye grow to large sizes, over one hundred kilograms, and generally seem to live no more than four to five years. The skipjack, although quite similar in many respects, does not grow as large as the other two; individuals greater than fifteen kilograms are relatively rare.

The temperate species, which include the albacore and bluefin, attain quite large sizes but their rate of growth is slower. There are from 7 to 15+ age classes in the fishery at a single time whereas the tropical tuna fishery is supported by far fewer age classes, say 2 to 5+.

All five species are highly mobile, some more than others, however. The albacore and northern bluefin make transoceanic migrations, and at various stages in their lives are caught in the eastern and western extremes of both the Atlantic and Pacific oceans. In the Pacific Ocean, skipjack are known to migrate from the coastal waters of the Americas to the central Pacific waters. The yellowfin and bigeye, though highly mobile, do not appear to make such extensive migrations as the other species.

The southern bluefin, which at certain times during its life occurs in subarctic waters, can be considered circumpolar since these fish migrate among the Pacific, Indian, and Atlantic oceans.

The Gear

By far the major share of all tuna taken in the world is captured using one of three methods. The first is longline, which consists of long strings of hooks suspended from a main line. The hooks extend vertically in the water column and reach depths of nearly one hundred fathoms. A single vessel generally handles approximately two thousand hooks which hang from a main line that may be as long as seventy-five miles. This form of fishing is used primarily by the Japanese and exclusively by the

Taiwanese and Koreans. Nearly 50 percent of the total catch of the five major species is made by longline.

The second most important fishing technique used to capture tuna is live-bait fishing. This technique involves chumming the tuna near the vessel with live-bait. When the tuna are feeding frantically, they are jerked from the water with poles to which artificial lures are attached.

The third most important fishing technique, in terms of total quantity landed, is purse-seining. This technique relies on the capture of surface schools of tunas with very large encircling nets. It is by far the most productive method of tuna fishing in terms of yield per day on the fishing grounds and is fast supplanting bait fishing as the most important surface fishing technique.

Other less productive techniques such as trolling with lures and fishing with traps account for nearly all of the remainder of the catch.

Tuna fishing is recorded in ancient times but did not begin in a significant way until after World War I and in a substantial way until after World War II. The United States fishermen fishing from California ports developed the live-bait tuna fishery in the eastern Pacific in about 1915, and by 1945 were fishing throughout the tropical Pacific waters as far south as Ecuador. In the western Pacific the Japanese live-bait tuna fishery extended from the home islands throughout the Caroline Islands. Purse-seining techniques, though developed prior to World War II, were not perfected until after 1958, when technological progress permitted the manufacture of materials and gear which were well suited to this form of fishing. Because of its great efficiency this fishing method has rapidly expanded throughout the Pacific and Atlantic oceans.

Longline fishing for tunas was developed by the Japanese prior to World War II, but was not used extensively until after 1950. As noted above this form of gear accounts for the major share of the world's catch of tuna. Longline vessels fish in nearly all tropical and temperate areas of the Atlantic, Pacific, and Indian oceans. Until about 1960, Japan was the only nation using this gear for tunas. They dominated the world catch, virtually controlled the markets, and their vessels were seen on all the major seas of the world. In recent years the Republic of Korea and the Republic of China (Taiwan) have entered the longline fishery for tunas and are fast competing with Japan for the dominant position in that fishery.

The Catches

Prior to World War II the total world catch of the five major species of tuna was about 300,000 metric tons (m.t.). After World War II it began to increase substantially and by 1952 was nearly 450,000 m.t. (Fig. 2). The catch increased steadily until about 1961 when it reached a plateau at about 900,000 m.t. and fluctuated about that level until 1965. Thereafter it continued to increase slightly, and in 1969 the catch of these five species amounted to approximately 1,100,000 m.t. (FAO, 1970).

Of this total catch the following species contributed in descending order of importance (Fig. 3): yellowfin 32.5 percent, skipjack 25.3 percent, albacore 20.0 percent, bigeye 12.5 percent, and bluefin 9.6 percent.

With the exception of skipjack in the Indian Ocean all of these species have been taken in significant quantities in the Atlantic, Pacific, and Indian oceans. Figure 4 shows the distribution of catch, by species, within oceans. It is readily apparent that by far the major catch of all species is taken in the Pacific. During 1969 approximately 63 percent of the total was taken there. The Atlantic was second, producing about 21 percent, followed by the Indian Ocean with the remaining 16 percent.

Just as the tunas are ubiquitous, occurring in nearly all of the tropical and temperate waters of the world, the nations which capture them are widespread and comprise a significant share of all the maritime nations of the world. During 1969 about forty nations reported capturing tuna. Though this is a rather impressive number, the statistics are misleading, for only six nations accounted for approximately 90 percent of the total catch, the other thirty-four nations accounting for the remainder. Of these six, Japan and the United States took nearly 70 percent of the world catch. Japan, by far the leader in the tuna fishery, captured about 50 percent (see Table 6).

TABLE 6

TUNA CATCHES OF PRINCIPAL PRODUCERS DURING 1969

Country	Catch Thousands m.t.	Percent of Total
Japan	537.0	47.6
United States	205.8	19.4
Republic of China	80.1	7.2
Republic of Korea	68.7	6.2
France	45.5	4.3
Spain	32.8	3.1
Subtotal	970.9	87.8
Others	129.1	12.2
Total	1,100.0	100.0

The Market

Tuna is an expensive commodity and in many less developed countries is too costly to compete with other cheaper products as a source of pro-

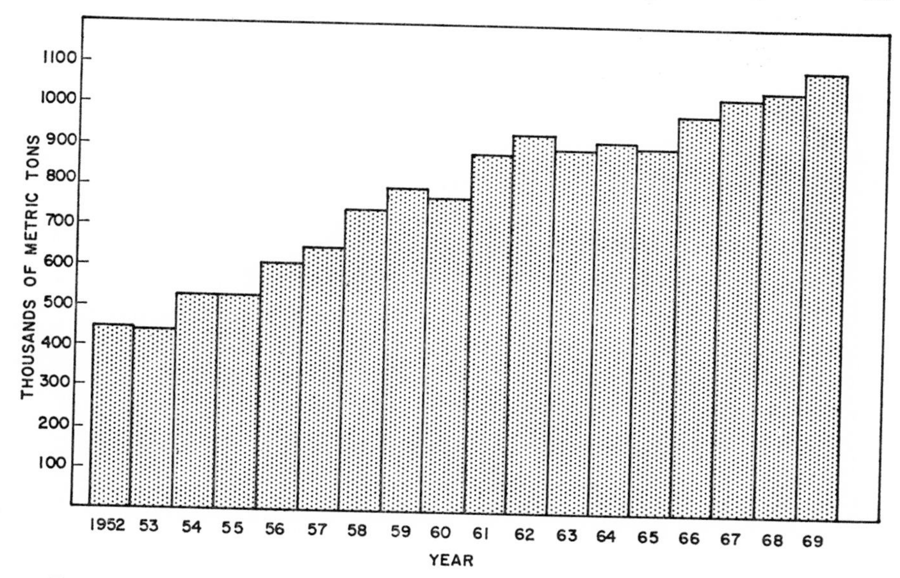

Figure 2. Annual world catch of the six major species of tuna, 1952–69

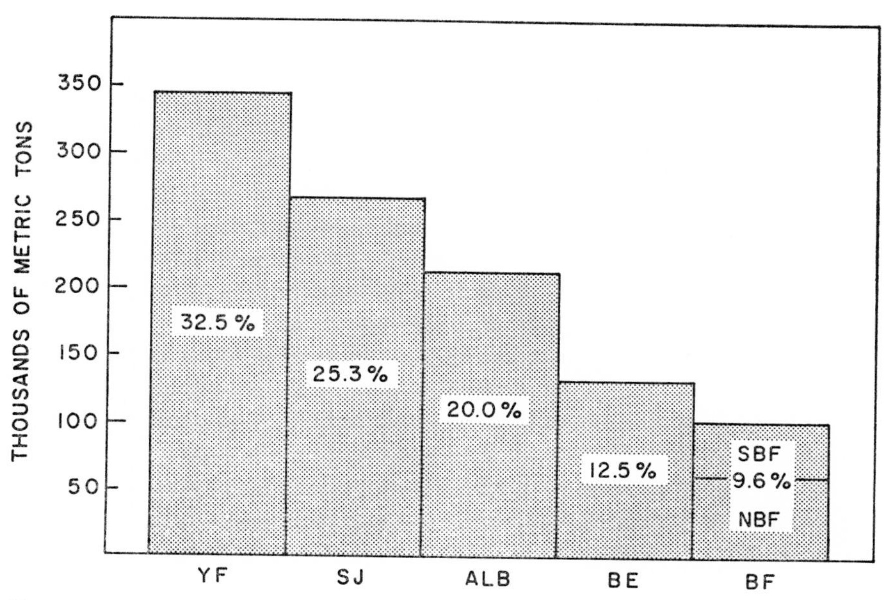

Figure 3. World catch of tuna, by species, for 1969. Numbers represent percent of total. YF = yellowfin; SJ = skipjack; ALB = albacore; BE = bigeye; BF = bluefin; SBF = southern bluefin; NBF = northern bluefin.

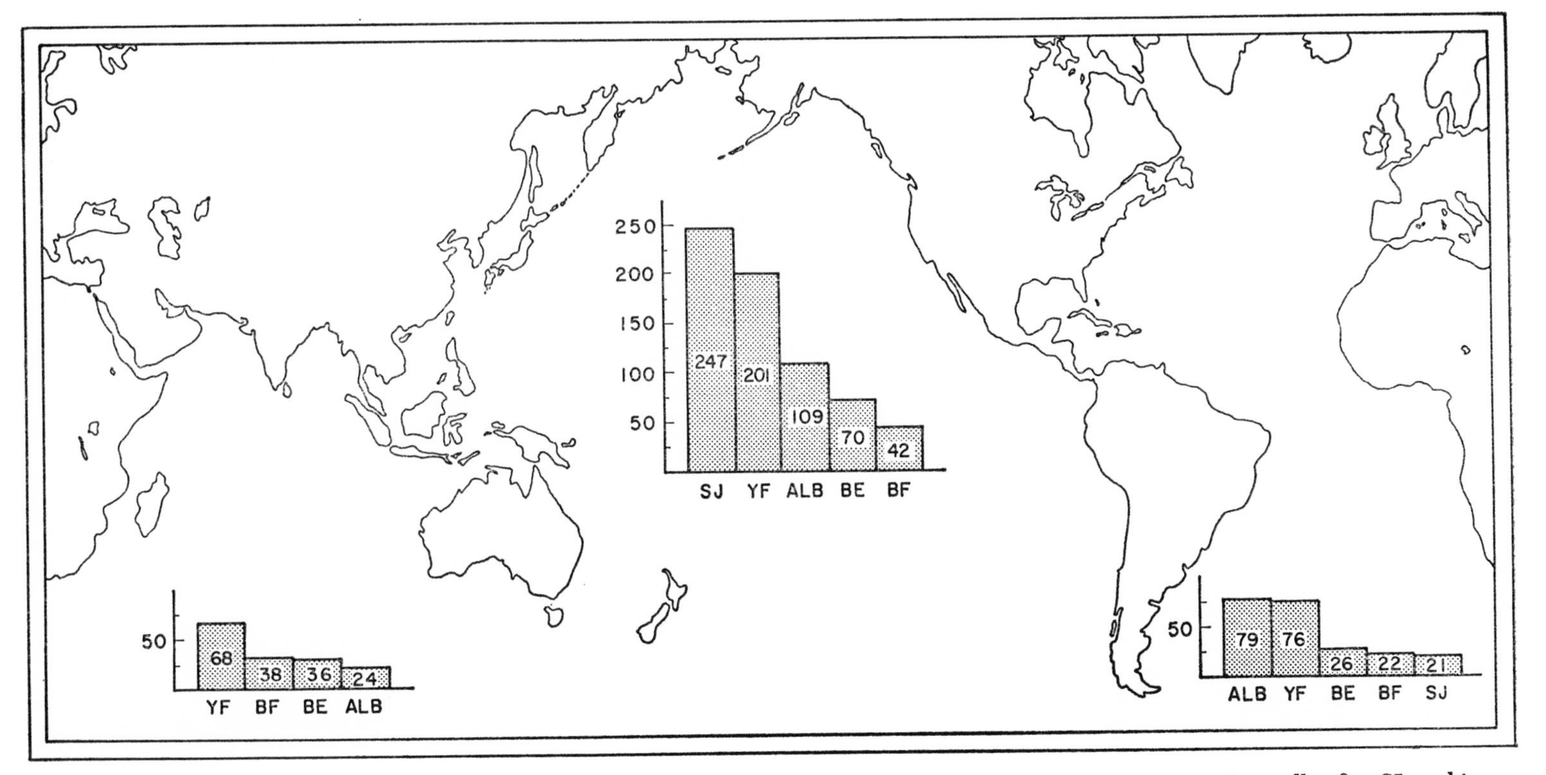

Figure 4. World catch of tuna, by species, by ocean, for 1969. Numbers represent thousands of metric tons. YF = yellowfin; SJ = skipjack; ALB = albacore; BE = bigeye; BF = bluefin.

tein. It is used by these countries as a source of foreign exchange. If we examine the utilization of the 1969 catch of the five major species we see that Japan and the United States utilize about 70 percent of the total catch, about the same amount as their combined catch (Broadhead, 1971). However, instead of utilizing the approximately 50 percent she captures, Japan utilizes only about 30 percent, while the United States, which catches about 20 percent, utilizes nearly 42 percent of the world catch. The developed countries of western Europe utilize about 19 percent while the remaining countries of the world use about 9 percent. Though over two thirds of all the tuna produced is captured and utilized by two countries, it is certainly an international fishery with a world market. Tuna is caught, shipped, and sold in all parts of the world. The price, taking into account labor and transportation costs, is rather comparable throughout the world and is not controlled by one nation. The total value of the world catch is significant: for example, during 1969 it was worth approximately $500,000,000 in equivalent United States dockside prices.

Condition of the Resource

During the fifties and sixties the production of tuna from the world's oceans has increased substantially. From 1950 to 1960 it approximately doubled, and during the subsequent years increased by nearly 25 percent. In the United States alone, the major tuna-consuming nation, consumption of tuna has nearly doubled every decade over the past fifty years (Chapman, 1967). This demand is expected to continue in the United States, as it is in other nations. The critical issue of course is whether the resources of tuna are capable of filling this increasing need for the raw product. Obviously we cannot expect growth in production as was experienced during the fifties since these were the years of great expansion. Can we expect any growth at all, or indeed is there some possibility that catches may decline as a result of overfishing?

During the late fifties and early sixties these questions were the subject of a great deal of concern. The Food and Agricultural Organization of the United Nations convened in 1962 a world meeting on tunas and related species. The purpose of this meeting was to call together experts from around the world to review the biology and present status of the world fisheries for tuna. This meeting resulted in the establishment of a permanent panel of a group of experts to facilitate research on the biology and fisheries of tunas and related species (FAO Expert Panel for the Facilitation of Tuna Research). From time to time this panel has established working groups to examine problems of particular importance related to the tuna fisheries. In 1968 it convened such a working group to examine the then current status of the tuna stocks in the Atlantic Ocean. Though the terms of reference of this group included only the Atlantic, they extended their analysis to include the Indian and Pacific oceans as well. In their report (FAO, 1968) the group deals with the five major species of tuna dealt with in this paper. A second report (FAO, 1969) by a similar group of experts, but dealing only with the Indian Ocean, updated the earlier one for that area.

In the next few paragraphs a very brief summary of the reports of these groups will be given, augmented by additional information.

Yellowfin

As noted earlier yellowfin comprise the major share of tunas landed in the world. They are captured in the Atlantic, Pacific, and Indian oceans. In the Indian Ocean they are taken by longline fishing techniques only. After examining statistics of catch and effort the above mentioned reports concluded that increased fishing effort generated in the longline fishery will not result in increased catches on a sustained basis, but in fact may result in a reduction of catch. This has also been concluded for the longline fisheries for yellowfin in the Atlantic and western Pacific. In the eastern Pacific, where a major surface fishery for yellowfin tuna exists, evidence demonstrates that the yellowfin stock was previously overexploited. A management program has been in effect in that area since 1966 which has resulted in maintaining the stock near a level where production can, on the average, be maintained at a maximum (Joseph, 1970). Increased production from the eastern Pacific is not expected, nor from the longline fisheries in other areas. However, our knowledge of the dynamics and biology of the yellowfin tuna indicates that increased production could be attained by harvesting the stocks of yellowfin at a smaller average size than are presently taken in the longline fishery. This assumes of course that it is possible to alter the average size of the fish in the catch, which is not certain. In any event, the maximum increase in yellowfin which might be expected from this would not be great and may only be from 5 to 20 percent.

Albacore

Albacore are taken in significant quantities in the three oceans and are captured by both longline and surface gear. The expert groups concluded that the albacore stocks of the world are presently fully exploited and increased effort on this species would not result in increased catches, but, again, could result in decreased catches. Whether production could be increased by altering the size structure of the catch was not evaluated, but knowledge of the life history of this species suggests that it probably could not be.

Bigeye

The fisheries for bigeye tuna, which take place in the Atlantic, Pacific, and Indian oceans, are essentially single gear fisheries. Nearly all bigeye are taken by the longline fishing technique. After an analysis of these fisheries, the expert group concluded that the bigeye stocks in the Atlantic and Indian oceans were probably not fully exploited but were nearly so, and that some slight increase in total catch might be expected from an increase in effort. They did not offer an opinion on the status of the bigeye stock in the Pacific, but as this fishery appears to be more heavily fished than those in the other oceans, very large increases in catch would not be expected.

Bluefin

As we have commented earlier, two species of bluefin occur in the world's oceans, the northern bluefin, which is confined nearly exclusively to the northern hemisphere and occurs in the Atlantic and Pacific, and the southern bluefin, which is found only in the southern hemisphere and is circumpolar.

The stocks of northern bluefin do not appear to be large, and those in the Atlantic appear to be heavily exploited. It was concluded by the expert group that increased effort on these stocks would most likely result in decreased catches. The catches of northern bluefin in the Pacific are also relatively small, but the potential for increase is unknown. The life history of this species and its relative abundance in the Atlantic Ocean indicate that the potential for increase would not appear to be much, if any.

The southern bluefin tuna, which apparently spawns in the eastern Indian Ocean and is captured in the Atlantic, Pacific, and Indian oceans, appears to be a single subpopulation. This species, which is by far the most commercially valuable tuna in terms of price per ton, is exploited almost exclusively by the Japanese longline fleet. Catches have declined from a peak of about 60,000 m.t. in 1960 to about 40,000 m.t. in recent years. During the same period, effort increased by about 50 percent. It has been suggested on the basis of these data and additional biological data that catches will not increase with additional effort, and most likely will decrease further unless effort on this stock is reduced.

Skipjack

Skipjack are found in the Atlantic, Pacific, and Indian oceans. This species comprises the largest component of the catch of tunas in the Pacific. In the Atlantic the catches are small relative to the other species of tuna taken, and in the Indian Ocean there is no major fishery for skipjack. Biological information suggests that this species is quite abundant, relative to the other major species, and in all oceans they appear to be underexploited. It is generally accepted that skipjack catches can be increased over the present levels. How much is not known, but for the eastern Pacific estimates of potential increase have ranged between two- and ten-fold (Rothschild, 1965; Joseph and Calkins, 1969).

To summarize this information on the condition of the resource, it is apparent that no great increase can be expected in the catches of four of the five species of tuna dealt with in this report. In fact in some instances it is possible that catches may actually decline unless effort is stabilized or reduced. The fifth species, skipjack, appears to be underexploited throughout the extent of its distribution. At a minimum, present catches can most likely be doubled.

Types of Existing International Organizations for Tuna Management

This brief review on the condition of the tuna resources points to the desirability of establishing some sort of rational regime of exploitation to prevent the reduction of the stocks at a level below that at which they are able to support their maximum potential yield on a sustained basis. The concept of such a regime, with a basis in conservation to prevent overfishing, is not a new one, dating back to the thirteenth century. Chapman (1970a) has reviewed the evolution of such systems.

There are two international organizations whose sole function is to provide the necessary guidance for the establishment of controls on the exploitation of tuna and related species. Though the purpose of these organizations is similar, their structure and operation are rather different.

Inter-American Tropical Tuna Commission

The first of these is the Inter-American Tropical Tuna Commission (IATTC), which was established by convention between Costa Rica and the United States of America in 1949. It is an open-ended convention in which any state whose nationals participate in the fisheries covered by this convention, providing acceptance of all member governments, may join. Five countries, Panama, Ecuador, Mexico, Canada, and Japan, adhered to the convention in subsequent years.* The convention waters are referred to as the eastern Pacific Ocean but are undefined in terms of longitude and latitude. The convention states that the commission's investigations shall cover not only the yellowfin and skipjack tuna but other kinds of fish taken by tuna vessels fishing in the eastern Pacific.

The IATTC consists of national sections, each comprising one to four commissioners, appointed by the member governments. All decisions,

* Ecuador withdrew August 21, 1968.

resolutions, and recommendations and other official action of the commission must be taken by a unanimous vote of all the high contracting parties. The IATTC elects annually a chairman and secretary to preside over meetings which are held at least once each year. The IATTC is empowered to designate a director of investigations who is responsible for the appointment and immediate direction of a scientific and technical staff to carry out the functions of the commission.

Most important among the duties of the IATTC is the responsibility to make recommendations, on the basis of scientific investigations, for proposals for joint action by the high contracting parties designed to keep the population of fishes covered by its convention at those levels of abundance which will permit the maximum catch on a sustained basis.

International Commission for the Conservation of Atlantic Tunas

The second of these international bodies is the International Commission for the Conservation of Atlantic Tunas (ICCAT). A convention for the establishment of this commission was drafted and signed in Rio de Janeiro in 1966. Seven ratifications were necessary to bring the convention into force, and the seventh country ratified during 1969.

The ICCAT convention is open for signature by any government which is a member of the United Nations or any of its specialized agencies. At the present time there are eleven members of the ICCAT: Spain, France, Brazil, Canada, the United States, Japan, Morocco, Portugal, South Africa, Ghana, and the Republic of Korea.

The convention waters comprise all waters of the Atlantic Ocean, including the adjacent seas.

The commission has responsibility for the study of the populations of tuna and tuna-like fishes (the Scombriformes with the exception of the families Trichiuridae and Gempylidae, and the genus *Scomber*) and such other species of fishes exploited in tuna fishing in the convention area which are not under investigation by another international organization.

Each of the contracting parties is represented in the commission by not more than three delegates, who may be assisted by experts and advisors.

The convention provides that, except as may be otherwise indicated, decisions of the commission are to be made by a majority of the contracting parties and that two thirds of the contracting parties constitute a quorum. Regular meetings are provided for once every two years.

The convention of ICCAT provides for research to be accomplished through three avenues: (1) utilization of the technical and scientific services of official agencies of the contracting parties and their subdivisions, and/or (2) utilization of the available services and information of any public or private institution, organization, or indivdual, and/or (3) independent research (within the limits of its budget) not accomplished under (1) and/or (2) above.

To carry out the objectives of its convention ICCAT may establish panels on the basis of species, groups of species, or geographic areas. At the present time the commission has established four panels: (1) tropical tunas—yellowfin and skipjack; (2) temperate tunas (north)—bluefin and albacore in the northern hemisphere; (3) temperate tunas (south)—bluefin and albacore in the southern hemisphere; (4) other species—bigeye, bonito, billfish, and others. Such panels shall be responsible for monitoring the populations of fish under their purview and the collection of information to do so, and for proposing to the commission, on the basis of scientific evidence, recommendations for joint action by the contracting parties. The commission, in turn, on the basis of scientific evidence, may make recommendations to the contracting parties, designed to maintain the populations of tuna and tuna-like fishes that may be taken in the convention area at levels which will permit the maximum sustainable catch.

There are two additional international organizations concerned with tuna whose terms of reference include research, development, and management of all species of fish within their geographical area of competence. The structure of these organizations is much different than the two tuna commissions in that they are established within the framework of the FAO.

Indian Ocean Fishery Commission

The first of these is the Indian Ocean Fishery Commission (IOFC) which was established by the Council of FAO under Article VI-1 of the FAO constitution on the recommendation of FAO's Committee on Fisheries (COFI). At the present time there are twenty-eight members.

The IOFC's geographical area of responsibility is designated as the Indian Ocean and adjacent seas but excluding the Antarctic area; the species of living marine resources with which it is concerned are not limited.

The objectives of IOFC are: (1) to promote, assist, and co-ordinate national programs over the entire field of fishery development and conservation; (2) to promote research and development activities in the area through international sources, and in particular international aid programs; (3) to examine management problems with particular reference, because of the need to take urgent action, to those relating to the management of offshore resources.

Membership in the IOFC is open to all member nations and associate members of the FAO.

The commission elects a chairman and up to six vice-chairmen at the end of each biennial session, but is not structured to support a research staff of its own. To accomplish its tasks it may establish any subsidiary bodies that it deems necessary, as well as call upon any outside expertise. With reference to the problems of conservation in the Indian Ocean the IOFC has established a committee on management which is charged with recommending, from time to time, on the basis of scientific evidence, measures for joint action by the member nations which are designed to prevent overfishing. This committee has in turn established a working party on stock assessment, in relation to immediate problems of management in the Indian Ocean, which has confined its activities to tuna.

Indo-Pacific Fisheries Council

The second of these latter types of international bodies is the Indo-Pacific Fisheries Council (IPFC). This body, which was formed within the framework of the FAO under the provision of Article XIV of the FAO constitution, was established in 1948. Its membership is open to all members and associate members of FAO, and such nonmembers as are members of the United Nations. Its areas of concern comprise both the marine and fresh waters of the Indo-Pacific region, and the species with which it is concerned are undefined.

The functions of the IPFC are broadly defined and include the following: (1) to formulate technical aspects of the problems of development and proper utilization of living aquatic resources; (2) to coordinate and encourage research and disseminate the results therefrom; (3) to recommend and undertake development projects within its member nations; (4) to propose and adopt measures to bring about standardization of scientific equipment, techniques, and nomenclature.

A chairman and vice-chairman are elected at each regular session of the IPFC but no provision is made for a research or technical staff. To accomplish its aims, the IPFC may establish committees and working parties wherever appropriate.

With reference to tuna the IPFC has established a committee on tuna management in the Indo-Pacific region, but at present it is more specifically concerned with the western Pacific.

Many types of arrangements other than international conventions and agreements have been made to manage fisheries resources, ranging from bilateral executive agreements to completely national programs. Obviously, national programs dealing with resources which spend their entire life cycle within the territorial waters of that nation are most advantageous. However, species of highly migratory animals which spend only a portion of their life within the territorial waters of the coastal state cannot be managed by the coastal state alone because controls which that state applies to such resources when in the waters under their jurisdiction do not apply when the animals are on the high seas.

The most effective means for managing common-property, high-seas resources in the past has been through the establishment of international conventions. Though many of these have so far been ineffectual in establishing the scientific basis for management and the subsequent recommendation and implementation of such, they provide the only examples of moderately successful high seas management programs in existence at the present time. Some of the more commonly known of these are: (1) the International Pacific Salmon Fisheries Commission (IPSC) whose management measures have been quite effective at maintaining the runs of sockeye and pink salmon into the Fraser River at optimum levels; (2) the North Pacific Fur Seal Commission (NPFSC); (3) the International Pacific Halibut Commission (IPHC); (4) the IATTC whose responsibility is tuna and tuna-like fishes in the eastern Pacific, and which since 1966 has maintained a conservation program for yellowfin tuna in that area; and (5) the International Whaling Commission (IWC). In some of these cases the management programs to date have been effective; in others they have not.

In a recent discussion of international fishery bodies Chapman (1970b: 337) has stated:

> Examples of major overfishing in individual nations (including the United States) can be cited that are worse than those in the high seas, and that leave one with the feeling that these problems are generally handled better, and quicker, in the international area where there is criticism from outsiders than they are when they are totally within national jurisdiction.
>
> The surprising thing is that, despite the acrimony that often develops in these meetings and activities of regulatory international fisheries commissions, problems get worked out, none of the commissions has ever gone out of business because of abrogation of the establishing convention, nations who resign have come back or continued to cooperate as well as they did when members, and non-members have frequently (if not ordinarily) abided by regulations that were established.

And further (Chapman, 1970b: 339):

> . . . the present system of independent fishery commissions composed of nations whose fishermen operate in a particularly high seas fisheries, operating under convention among them, works slowly and often badly, but the system does work and can prevent overfishing of high seas resources that is of serious, or permanent, consequence. It can, if pressure is strong enough, also compose the economic differences between the nations that arise from the conservation regulations that ensue.

During the last two decades there has been a proliferation of fisheries commissions and international organizations established, which would suggest that such bodies offer the best means available at the current time for dealing with fisheries conservation problems in the international arena.

Assuming that international commissions provide a desirable and functional basis for managing the high-seas tuna resources in the oceans of the world, then how can they be structured and operated to attain their objectives in the most expedient manner? This problem can be divided into two general categories. The first includes the structure, terms of reference, and general operation of the commission itself; the second is concerned with the manner in which the recommendations for joint management action are formulated. The remainder of this paper will deal with these two items.

A Comparison of Present Arrangements and Possible Modifications

Structure

The four international fisheries bodies (IATTC, ICCAT, IOFC, and IPFC) discussed above have at least one common objective among them, that is, the conservation of tunas and tuna-like fishes within their areas of geographic responsibility. How these organizations are structured to attain these objectives is quite different, as we have already seen.

Two of them, the IATTC and the ICCAT, are independent international commissions established by international treaties and as such are answerable to the states signatory to the treaty. The other two, IPFC and IOFC, are established by agreement within the framework of the FAO and as such are not independent from that organization. Indeed, these two latter bodies are not distinct from the FAO and, as regional subsidiary bodies, remain subject to the control of the FAO conference and council, and hence their activities are carried out through FAO (Carroz and Roche, 1967).

However, the ICCAT, in contrast to the IATTC, is not completely independent of the FAO either. In a review of the functions and structure of ICCAT and its relation to other bodies, Carroz and Roche (1967: 700–1) state:

> In view of the special responsibilities of FAO with respect to fisheries development, the Atlantic Tuna Convention also provides that the Commission shall enter into negotiations with FAO regarding the conclusion of a relationship agreement which should provide for the appointment by FAO of a representative who would participate, without the right to vote, in all meetings of the Commission and its subsidiary bodies.

No such arrangement, which specifically singles out the FAO as a participating body, is provided for in the treaty establishing the IATTC. The need for coordinating the activities of the various bodies concerned with tuna management is certainly a strong one, but the method of doing so needs further examination and will be touched upon in a later section of this report.

All four of the bodies are similar in that their treaties and agreements contain no provision on their legal status nor do they specify their legal capacity under international law or national legal systems to perform their established duties. This is common to all intergovernmental fisheries bodies except one, the International Council for the Exploration of the Sea, which provides for such legal status and capacity in its convention. Though international fisheries bodies have not been granted legal status they have nevertheless been able to perform legal acts necessary to their objectives on an *ad hoc* basis, including contracts and transactions relating to the purchase of equipment, the hiring of staff, the chartering of vessels, and the leasing of premises. However, there is no basis upon which these organizations can enforce their rights, and in certain instances the lack of legal status has obstructed the timely performance of their duties. These obstructions have included, among others, the inability to enter into certain forms of legal contracts, to operate research vessels in the territorial waters of member states, and to effect the free entry and exit of scientific materials and equipment into and from the territories of member and cooperating states.

Even though these *ad hoc* arrangements have generally sufficed to permit the expeditious operation of these international bodies in the past, it would appear desirable that conventions creating such bodies include a provision establishing the legal capacity of such bodies to perform their designated duties.

Methods for funding the four bodies differ remarkably. The IATTC and the ICCAT are provided with funds to carry out the duties designated in their conventions. Funds are contributed by the member governments in a prescribed manner. Members of IATTC pay in proportion to the amount of fish, originating from within the convention waters, utilized by each member, regardless of who captured it. Contributions by member states of the ICCAT are determined on the basis of catch and utilization, and by membership on panels. The conventions of IOFC and IPFC do not provide for an operating budget; these commissions rely on the FAO for support.

Probably one of the most important and vigorously debated subjects concerned with the establishment of fisheries commissions is that of staff. Before management recommendations are made there must be demonstrated a need for such measures and the benefits which might accrue from them, based on scientific evidence. To attain such evidence requires research by trained scientists. How this research should be accomplished and under what political regime the scientists should operate has been a heated subject of debate.

There are two extreme points of view which can be taken on this subject and both are represented in the four examples we have discussed here. On the one hand there are commissions such as the IPSC, IPHC, and the IATTC which have a full scientific staff, under the supervision of a director of investigations, to carry out all of the necessary research of the commission, while at the other extreme are organizations such as the IPFC which have no staff whatsoever except an elected chairman and vice-chairman and who must rely on committees composed of member countries to do any technical work which may be necessary. In between these two extremes are commissions such as ICCAT and ICNAF which have a permanent secretariat to attend to certain fiscal and administrative matters, but which must rely on panels and committees composed of scientists from national sections to accomplish the research. There are a number of pros and cons for the establishment of a staff, some of which are particularly apropos to problems concerned with tuna. One obvious point is that if there is no national expertise in the fisheries sciences, then not much research of a significant scale can be accomplished by that nation. Many less developed countries (LDC's), some of which fall within the latitudinal zones in which tuna occur, have not had a very long history (and in some cases none whatsoever) of technological development and thus have not been able to develop trained scientists. This may be particularly true in fisheries, where the immediate problem of concern with the LDC's has been one of fisheries development and not fisheries management. These countries may not want to and perhaps should not rely upon the expertise of the developed nations who are members of the particular commission of concern. In many cases it seems natural to suppose that these developed nations by virtue of their technology and economy may be interested in research which is oriented toward completely different objectives than those of the LDC's. Research which is conducted on common property resources, under multi-nation exploitation, by scientists of national sections of an international body, may be influenced by the needs of the nation to which these scientists belong. Likewise the possibility always exists among the international fishery bodies which do not retain research staffs that seemingly minor differences in research results of national sections will be allowed to obscure what might be adequate scientific evidence for management decisions to be based upon. Though scientific research should be, and always strives to be, free of political and economic externalities, it is not always possible to keep them separated.

It appears intuitively obvious, however, that nations with advanced research capabilities in the field of fisheries science are reluctant to give up research programs to international organizations because this would reduce the number of options open to them for their own research programs and they would lose control over the direction and possible outcome of the research. Additionally, both LDC's and developed countries are reluctant to provide the financial outlay necessary to hire independent research staff for fisheries commissions.

Tuna, we have noted earlier, are captured worldwide. They are landed in numerous ports in the oceans of the world and enter international markets on a significant scale. Information on the fishing activities of the vessels in the fishery, fishing effort, catch rates, total catch, and so on, provides the basic data needed to study the dynamics of the fishery and to assess the effect that man has upon the abundance of the stocks by his fishing activities. The collection of such data is an onerous task requiring the effort of a diligent cadre of technicians who can maintain the confidential nature of such data. The commissions with independent staff have been particularly successful in this task.

In the past, only three commissions have been provided with a permanent, professional research staff. It is noteworthy that each of these has accomplished its designated duties. It is also noteworthy that since the establishment of the IPSC prior to World War II, only one fishery commission has been provided with an independent staff. When the question of an independent staff was raised during the formation of ICCAT, it was rejected. Burke (1967:178) had the following to say on this subject:

> The Atlantic Tuna Conference is perhaps instructive in view of the observation that "representatives of seventeen nations widely differing in interest, economic and political ideas, and in degree of development, all collaborated harmoniously to a common purpose." It is true that the delegates managed to produce a convention establishing an Atlantic Tuna Commission and in this sense harmony triumphed. However, the states involved conferred very little authority upon the Commission and did not even endow it with an independent staff. It is easy to produce harmony when the cost is low. The value of this experience is not likely to be great since future problems probably can be resolved only by establishing groups equipped with an appropriate staff and with sufficient authority to deal with the problems they face.

Because of the technological and economic development of the major share of the tuna fishing nations, the widespread distribution of the resources, the fleets, and the markets, the high-seas nature of the tuna themselves, and the success of fisheries commissions with independent research staffs, it appears that the purpose of the various fisheries commissions would be best served if they were endowed with independent research staffs.

The four bodies we have been discussing differ rather sharply regarding the geographic areas and species of fish for which they are responsible.

The IATTC convention waters are designated as the eastern Pacific Ocean and the species of concern are yellowfin and skipjack tuna and other kinds of fish taken by tuna fishing vessels in the eastern Pacific. The ICCAT convention is broader with respect to area, designating the Atlantic Ocean and adjacent seas, but with respect to the species is more specific, referring to tuna and tuna-like fishes. The IOFC includes all species of fish within their responsibilities which occur in the waters of the Indian Ocean and adjacent seas, but specifically excludes the Antarctic area. Finally the IPFC is most general in that it designates no species and refers to the Indo-Pacific area, both fresh and marine, as its area of responsibility. Thus we see the world oceans have been partitioned into regional areas for the purposes of tuna management. The obvious question to pose is whether this is the best way to approach this problem.

To evaluate this question we first must examine the animals which are themselves the object of management. Research has shown that tunas are highly migratory fish, some species more so than others.

Tagging experiments have demonstrated that albacore and northern bluefin tuna make transoceanic migrations. In one year albacore off the east coast of Japan migrate to the west coast of North America (Clemens, 1961). Northern bluefin tagged off the eastern United States have been recovered in northern European waters, in the Bay of Biscay, and off Brazil (FAO, 1968). Catch statistical information suggests the possibility that albacore and yellowfin found at one stage of their life in the southeastern Atlantic Ocean may occur later in the southwestern Indian Ocean (FAO, 1968). Skipjack tuna have been shown to migrate from within the extremes of the eastern Pacific Ocean to the central Pacific and indeed are only seasonal visitors to the eastern Pacific (Joseph and Calkins, 1969). Recent studies based on genetics of protein systems suggest that a single skipjack subpopulation occurs from the mainland of the Americas to about 150°E, and a second subpopulation from there to mainland China (Fujino, 1969). Tagging and other biological studies have demonstrated that southern bluefin tuna are a single, intermingling population which is distributed circumpolarly throughout the Atlantic, Indian, and Pacific oceans. Yellowfin tuna are more migratory than originally thought, according to recent tagging information (IATTC, 1971).

All of these examples demonstrate that tuna are highly mobile animals. The tunas themselves do not recognize imaginary boundaries and in order to manage them properly, the establishment of such boundaries is not realistic.

The problem of imaginary boundaries can also be extended to those established on the basis of territorial claims of the coastal states, whether these be 3, 6, 12, 80, 120, or 200 miles in breadth. The animals themselves do not recognize these boundaries and therefore tuna conservation programs established on that basis will not work. In no case will unilateral, or even multilateral action taken within territorial seas, whatever the breadth, suffice to allow the proper management of tuna species since tuna are too wide-ranging. Therefore, in formulating fisheries conventions which apply to tuna and other high-seas resources of a migratory nature, the convention area with respect to scientific research and management should include the territorial waters of the coastal states if the fish occur therein during some stage of their life.

Likewise the areas over which the research and management of a fisheries body extend should include the range of the fish under study. An example here is the IATTC whose convention waters are the eastern Pacific. It has already been demonstrated that for skipjack tuna certainly, and yellowfin tuna possibly, the westward extent of the convention waters, even though imprecisely defined, does not include the entire range or population of animals which fall within the responsibility of the IATTC. More specifically the southern bluefin tuna will present even greater problems if it becomes necessary to manage this species because they comprise a single subpopulation and occur within the convention waters of all four of the international bodies concerned with the management of tuna.

On the basis of the biology of the tunas there is good reason to suggest that geographical areas of responsibility need to be based more on biological parameters and less on political ones.

Not only are the fish themselves highly migratory but the fleets that capture them operate over vast areas. In a single year one vessel may fish in the Atlantic, Pacific, and Indian oceans; therefore when management decisions are made, this great fleet mobility must be taken into account.

Management Functions

Once a system has been formulated for conducting scientific research necessary to establish the levels of fishing required to prevent overexploitation, the next logical step is to implement this advice. There are a variety of experiences which can be referred to in this context. These include:

1. The Pacific halibut fishery, which is controlled by the IPHC composed of Canada and the United States on the basis of total catch divided among areas and seasons.

2. The Fraser River salmon fishery controlled by IPSC. The size of the catch quota is based on the escapement and is shared in equal parts by the two member nations, Canada and the United States.

3. The north Pacific fur seals, the harvest of which is controlled by the North Pacific Fur Seal Commission, composed of Japan, the USSR, the United States, and Canada. The number of seals to be harvested each year is established by the commission and taken by the USSR and the United States. The proceeds from these are proportioned to the member governments in a prescribed manner.

4. Certain species of whales in various sectors of the world, the harvest of which is controlled by the International Whaling Commission. The catch is determined by the number of blue whale equivalents which can be harvested, and is divided among countries in the form of quotas.

5. Yellowfin tuna of the eastern Pacific which is controlled by an overall catch quota established by IATTC.

Though each of these commissions has implemented management programs that have been successful to differing degrees, they have all been confronted with the continuing problem of allocation. There has been general agreement upon the amount of the resource that should be harvested, that is, the maximum sustainable physical yield, but within this limitation of maximum yield the overwhelming problem has been, and continues to be, *who gets the fish.* This problem of course has been an important one for centuries and undoubtedly will continue to be so. It was one of the major issues of the 1958 Law of the Sea Conference (LOS) and will be again at the LOS 1973.

To examine this problem of allocation of the resource in slightly greater detail, particularly for tuna, and its relationship to international fishery commissions, it is of value to review the experience of the IATTC. As noted earlier the main objective of the IATTC is to gather and interpret factual information to help maintain fish populations within the area covered by the convention at a level permitting maximum catches year after year. The tuna fishery in the convention area is based primarily on two species, yellowfin and skipjack. The commission's research showed at an early date that skipjack tuna were underexploited and that yellowfin tuna were being harvested near the level of maximum sustained yield. By 1961, with increasing fishing effort, the catches exceeded this maximum level and the commission, on the basis of scientific evidence, recommended that the catch be restricted to a level which would permit the stock to rebuild to its optimum level. The recommendation for management was in the form of a resolution requesting all countries participating in the fishery for tuna within the convention area to cooperate in the management program. This cooperation was necessary because not all states fishing in the convention waters were signatory to the convention. Additionally, the Tuna Conventions Act of the United States, the major producer and user of tuna from within the convention waters, states that regulations for the control of tuna fishing in the convention waters shall not be promulgated:

> . . . prior to an agreed date for the application by all countries whose vessels engage in fishing for species covered by the conventions in the regulatory area on a meaningful scale, in terms of effect upon the success of the conservation program, of effective measures for the implementation of the Commission's recommendations applicable to all vessels and persons subject to their respective jurisdiction. The Secretary of the Interior shall suspend at any time the application of any such regulation when . . . he determines that foreign fishing operations in the regulatory area are such as to constitute a serious threat to the achievements of the objectives of the Commission's recommendations.*

Therefore without the participation of the United States, there could be no effective conservation program, since this country takes the major share of the catch. The commission in its recommendations never went beyond the concept of an overall catch quota. Peru and Chile considered that such a program would not allow for the growth of their emergent tuna industries and insisted on some form of country allocation. This could not be agreed to within the commission, and since Peru, at least, captured significant quantities of tuna, regulations in turn could not be agreed to. By 1966, Peru's tuna industry had virtually disappeared because of her interest in the anchoveta reduction fishery, and since Peru was no longer a "serious threat to the achievements of the objectives of the Commission" as specified in the Tuna Conventions Act (Chile's catch of yellowfin tuna was also minimal), the recommendations of the commission were implemented for the first time in 1966.

The conservation program has been in effect since that time, although it has been difficult to maintain. There has been a continually expressed need on the part of some member states for some form of allocation among participating countries. Mexico has been particularly conspicuous in this regard, indicating that her developing fleet, which is composed primarily of small vessels, is handicapped by the present form of the tuna management program. She expressed her need for a new system which would allow for the development and growth of her emergent tuna industry. To allow for special problems of small vessels in the fishery, in 1969, the commission instituted, on the recommendations of an intergovernmental body composed of all countries participating in the

* Tuna Conventions Act of 1950, Public Law 764, 81st Cong., 2d sess., 64 Stat. 777, amended 15 October 1962, 76 Stat. 923, 16 U.S.C. 951–61, sec. 6(c).

fishery, special allocations amounting to four thousand tons per country for vessels under a minimum size. This was increased to six thousand tons in 1970, and was continued in 1971. Furthermore, a provision was made for 1971 to allow member countries which are developing a tuna fishing industry, which have vessels entering the fishery for the first time after the closure date to unrestricted fishing, to take an additional two thousand tons.

All of these special allocations were, in fact, an effort to compensate for the needs expressed by member states. Realizing these measures were of a "stop-gap" nature the commission established a working group to examine alternate methods of regulation, which would lead to a long-term solution to the management problem.

During the period since regulations were first implemented, conditions in the fishery have changed remarkably. More nations are participating in the fishery, fleet capacity has increased sharply, and the catch within the regulatory area has remained relatively stable. The international fleet, which had a capacity of forty thousand tons in 1962, now has a capacity of about seventy-five thousand tons, an 88 percent increase. Whereas vessels had an opportunity to fish for yellowfin tuna the year round prior to the implementation of the management program, the open season has progressively grown shorter (i.e., for the period 1966–70, 8½, 6, 5½, 3½, and 2½ months, respectively). Obviously, the form of management recommended by the commission has led to overcapitalization for yellowfin within the Commission Regulatory Area (CRA). However, because the tuna fleets are highly mobile and can fish areas other than just the convention waters, one needs to examine their activities in the area in light of their activities in other waters. As noted above, since 1962 the fleet has increased 88 percent. After examining the total catch of this fleet in 1962 and in 1970, which includes fish from the Atlantic and central Pacific as well as from within the CRA, one observes that the catch has increased by about 60 percent. Though the catch has not increased in proportion to the increase in fleet capacity, one does not get the same picture of gross overcapitalization when only activities within the CRA are considered .

The international fleet which operates in the CRA is expected to increase by about a third (to approximately one hundred thousand tons of capacity), within the next two years. Such an increase will indeed aggravate the problems of maintaining a conservation program in the eastern Pacific.

A number of solutions have been discussed. The concept most frequently considered has been that of some system of allocation in the form of country quotas. Obviously the problem here is one of the basic distribution of catch which presently exists in the fishery. Eleven countries now have vessels fishing in the CRA. The United States fleet, however, captures 82 percent of the total catch of yellowfin. The next most significant country in the fishery is Mexico, which takes about 6 percent of the total catch. Any division of the catch among countries would be done at the cost of decreased catches for the United States, whose producers, naturally, are reluctant to do this. If a system of country quotas were established within the convention area it is logical to assume that these would be negotiable as countries "grew into" their quotas. It would not be unreasonable to expect that as the United States share decreased, vessels would leave the fishery to assume the flags of other countries, to fish for a share of the quotas of those countries. (In fact, this strategy has already begun under the present system of special allocations.) If we include the countries bordering the fishing area together with countries presently capable of fishing in the eastern Pacific, we find that about twenty states might come within the quota system. Considering the maximum sustainable catch to be about 100,000 to 120,000 tons per year this results in about 4,000 to 6,000 tons per country. With an allocation of this size no country would be able to expand its industry beyond the current level of Mexico's tuna industry; and at the present level Mexico finds its tuna industry is not adequate to fill the needs of her people. It would appear then that such a system of country quotas would leave us no better off than the present system of free access, and in fact would perhaps be even less desirable.

If country quotas are to be considered for the eastern Pacific fishery, then they must be considered in light of a time-horizon over the next few decades. Though they appear perhaps desirable on the short term, what is their effect over the long term? It is not difficult to predict the existence of tuna conservation programs in the Atlantic, Indian, and western Pacific oceans in the relatively near future. Any system of regulation, especially country quotas, must take into account the possibility of regulations in other oceans and their effect on any possible programs in the eastern Pacific and vice versa.

Because there are generally considered to be more vessels in the CRA than are needed to take the allowable harvest, various forms of limited entry have been considered. It is difficult, however, to conceive of limited entry without some underlying system of allocating the catch among the participants. Therefore such allocation appears to be a necessary antecedent of limited entry. To cite an example of why this is so one needs only look at the Japanese longline fishery for tunas and related species. The Japanese maintained a system of licensing entry into the longline fishery to keep the catch and catch rate at a level which would insure a profitable fishery. This worked quite well until the Korean and Taiwanese fleets, who have not limited their own entry, entered the fishery.

The result has been that the Japanese, who are forced to operate on lower catch-rates, and hence on lower profits, are being driven out of the longline fishing business.

An additional alternative system which has been suggested, but never examined in much detail with regards to tuna (but ought to be because of the wide interest shown in the concept in recent years), is the possibility of granting some management body very much broader powers over common property resources than any international body has presently. Numerous articles have appeared in the literature discussing this concept; the reader is referred to Christy and Scott (1965) and Gulland and Carroz (1968) for further reading on the subject. Their system would give control of access to the resource to some international body which might use a system of taxing, licensing, or competitive bidding to grant access. The proceeds from such action could then be used to finance the research and enforcement activities of the international body and/or could be redistributed among the community of nations or member governments on some system of pre-established criteria.

For such a system to operate in the eastern Pacific fishery the participating states, of course, must agree to forfeit their present right of free access to the resource. Bidding would be based on the economic status of the various national fleets and on the designated national programs of development among the countries. Certainly the strategy of bidding could be influenced by the options open to the participants in tuna fisheries in other ocean areas. Therefore, once again in consideration of such a system, predictions must be made of future arrangements in other ocean areas and their effects on fishing strategy within the eastern Pacific.

A Suggested Future Arrangement

It seems obvious that no one has been able thus far to devise an alternative management program for the yellowfin tuna fishery of the eastern Pacific which is responsive to the needs of the developing tuna fishing nations, meets the needs of those nations which are already highly developed tuna fishing nations, allows for control of fishing on yellowfin tuna but yet encourages increased harvests of skipjack and other underfished tuna-like fishes, takes into account the present and potential tuna fisheries in areas other than in the eastern Pacific, and is logistically manageable. It appears that the present system will not withstand the pressures of the participating countries for change and therefore the conservation program will be endangered. Similarly, when management programs are recommended by the other commissions for the areas of their responsibility, problems similar to those in the eastern Pacific will occur. Therefore a management strategy is needed that will be adequate to handle these problems, and a mechanism is required for developing such a strategy.

What then can and should be done to develop such a strategy and to establish such a mechanism? Before an attempt is made to answer this, several points presented earlier are re-emphasized here because of their relevance to the discussion below.

1. The total world catch of tuna (1.1 million m.t.) is taken by approximately forty countries. Of these, two countries, Japan and the United States, take nearly 70 percent and also utilize about the same amount. Six of the forty countries account for nearly 88 percent of the total, the remaining thirty-four countries taking the remaining 12 percent of catch.

2. Tuna are wide ranging, highly migratory fish. They make transoceanic migrations and even migrate among oceans. They do not recognize imaginary boundaries established by man.

3. The fleets that capture tuna are also highly mobile; individual vessels can, and in fact some do, fish tuna in all of the world oceans.

Any management program considered for any part of the world must take into account these facts, as well as numerous others. Such programs cannot be considered on a regional basis only, nor can they consider biological factors alone. They must be receptive to the needs of nations, industries, and individuals. Fishery management can no longer tend to take a narrow view of its problems. In fact, Rothschild (1971) has discussed this problem in some detail and has suggested that we have not adequately recognized the need to develop techniques to handle these complex problems of fisheries management. In his paper, Rothschild discussed approaches which involve systematic techniques for studying such problems on a broad and comprehensive basis. Obviously such an approach is necessary to develop management programs for tuna which take into consideration the complexity of the fishery.

To accomplish this we see that an analysis of the eastern Pacific fishery cannot be taken as an entity in itself, but must include all of the tuna fisheries of the world. It would therefore appear that the concept of regional commissions for tuna will be inadequate. Though such regional commissions have served us well to date, and the establishment of fishery commissions has certainly been successful, the regional commissions concerned with tuna need to be expanded in scope. Not only is this true with respect to the research aims of the commissions but it is especially true for the management aims. There is a strong need to pull the international bodies concerned with tuna closer together. Two of the bodies, the IOFC and IPFC, are within the framework of FAO and as such their activities can be easily coordinated. The ICCAT, though independent, has a close working relationship with FAO. IATTC has no relationship whatsoever with FAO and hence none, other than on an *ad hoc* basis, with the other bodies. If these bodies retain their present form some concrete manner of contact needs to be established between the two. The contact must be of a form which can influence the decision-making processes in each of the bodies. This must be so because of the nature of the fish, the fleet, and the market, as listed above; management decisions made in the eastern Pacific will affect fishery strategy in other oceans and hence management decisions there. This coordination among regional bodies could, at a minimum, be accomplished through the establishment of an *ad hoc* advisory group or a formal committee established within the framework of the FAO. An alternate approach would be for those countries who are signatory to more than one convention to assign the same commissioners to each one.

A more straightforward approach, and one that in my opinion appears more efficient, has been presented by Kask (1968), which calls for the establishment of a world tuna commission, or some such similar body. He based his opinion, as I have done, on the international nature of the resource, industry, and market, and on the recommendations presented in the report of the FAO's (1968) Expert Panel for the Facilitation of Tuna Research on the assessment of the tuna stocks of the Atlantic Ocean. This report stated in part (p. 35):

> There is an urgent need for an improvement in the statistics of total landings, species composition, and fishing effort. Because of the nature of the fisheries—long-range vessels and vessels landing in foreign countries—the collection, tabulation and publication of detailed statistics might be better done for the world as a whole, rather than for each ocean separately.

The report also states (p. 31):

> The Pacific tuna fishery is based on the same species, is largely carried out by the same countries (and indeed, often the same vessels), and supplies the same market as the Atlantic and Indian Ocean Fisheries. It is therefore unrealistic to consider any one of these oceans in isolation as regards statistics, scientific research or management.

If a world tuna commission is called for, how then should it be structured, what should be its terms of reference, how should it be financed, and what should its legal status be? Kask, in the above-cited reference, has addressed himself to some of these problems and they have been discussed in more general terms earlier in this paper.

The commission should be established by convention and given legal status as an independent entity so that any future contingencies might be met in a straightforward manner.

The terms of reference should be simple and precise, but not so restrictive that problems of a significant nature cannot be studied. The objectives of the commission with respect to conservation should not be too restrictive, for example, the attainment of maximum physical yield as with the present IATTC and ICCAT. The commission should be free to examine alternate forms of management which might be optimized through consideration of economic, social, and other factors.

The convention waters should not be restrictive, but should include all waters in which occur the species with which the commission is concerned.

The species to be covered should include all species of tuna, bonito, and billfishes listed in the FAO *Yearbook of Fishery Statistics* which spent at least some part of their life in the high seas.

The commission should have its own research staff, and should be provided with a budget adequate to carry out its duties.

Membership in the commission should include all nations that are engaged in the tuna fishery and that catch and/or process some predetermined minimal amount of tuna; membership should not be restricted to the FAO family of nations. A system of voting in which the vote is weighted by participation and investment in the fishery would be worthy of consideration.

Obviously the structuring of a commission along these lines which would be agreeable to the community of nations and which would not be too cumbersome to operate efficiently and effectively is a formidable task. From a scientific point of view this appears the logical path to follow. The task of determining whether such a scheme is even possible, of course, will fall upon the lawmakers and diplomats.

Summary

Tunas are a world resource occurring in all the tropical and temperate waters of the world. They are captured by an international fleet of vessels representing numerous nations. The product is sold in an international market.

The six major species of tuna are heavily fished and there is probably more vessel tonnage capacity available than is needed to harvest them, or at least there soon will be. Certain stocks are apparently either fully exploited or perhaps overexploited, others cannot produce much more than the present level, and only one of the major species is capable of supporting a significantly larger production.

There are four regional, international bodies in existence which are concerned with the research and management of the tuna stocks. Of these only one has a permanent research staff, and only one has conducted the necessary research to recommend management, that is, management of yellowfin tuna in the eastern Pacific Ocean.

The management program, based on a general quota system, has been in force since 1966 and has been successful so far. However, due to increased demand and competition for the raw product, the continued success and maintenance of the program is uncertain. The commission has been unable to develop a new system of management which meets the needs of the participating countries. Similar problems will appear in the other regional commissions if and when management is implemented.

With respect to international fisheries and the future problems related thereto, fisheries science has not been responsive to the complex nature of the problems.

Because of the complexity of the problems related to research and management, and the international nature of the resources of tuna and the industries connected therewith, it is suggested that the present concept of regional commissions is not adequate to deal with these problems. The establishment of a world tuna commission appears necessary to deal with these problems. This commission must be prepared not only to deal with problems of a scientific nature concerning the resources themselves, but must be responsive to the needs of the member nations in developing management strategies.

January 1971

REFERENCES

Broadhead, G. C. 1971. *International Trade-Tuna.* Food and Agriculture Organization, Indian Ocean Fishery Commission, IOFC/DEV No. 14. 27 pp.

Burke, W. T. 1967. "Aspects of International Decision-Making Processes in Intergovernmental Fishery Commissions." *Washington Law Review,* 43:115–78.

Carroz, J. E., and A. G. Roche. 1967. "The Proposed International Commission for the Conservation of Atlantic Tunas." *American Journal of International Law,* 61 (July): 673–702.

Chapman, W. M. 1967. "Recent Trends in World Tuna Production and Some Problems Arising Therefrom." *Proceedings of the Symposium on Scombroid Fishes,* Part III, pp. 1173–83. Mandapam Camp: Marine Biological Association of India.

———. 1970a. "The Theory and Practice and International Fishery Development-Management." *San Diego Law Review,* 7:408–54.

———. 1970b. "Some Problems and Prospects for the Harvest of Living Marine Resources to the Year 2000." In U.S. Congress, Senate, Committee on Commerce, Subcommittee on Oceanography, *Hearings on S. 2841 and S. 2802,* 91st Cong., 1st and 2d sess., pp. 320–49. Washington, D.C.: Government Printing Office.

———. 1970c. "Seafood and World Famine: Positive Approach." Papers presented at a symposium on food from the sea, American Association for the Advancement of Science, Annual Meeting, Boston, December.

Christy, F. T., Jr., and A. Scott. 1965. *The Common Wealth in Ocean Fisheries.* Baltimore, Md.: The Johns Hopkins Press for Resources for the Future.

Clemens, H. B. 1961. *The Migration, Age and Growth of Pacific Albacore* (Thunnus germo), *1951–1958.* California Department of Fish and Game, Fish Bulletin No. 115. 128 pp.

Food and Agriculture Organization. 1968. *Report of the Meeting of a Group of Experts on Tuna Stock Assessment* (August 1968, Miami). FAO Fisheries Reports, No. 61. 45 pp.

———. 1969. *Report of the IOFC Working Party on Stock Assessment in Relation to Immediate Problems of Management in the Indian Ocean* (September–October 1969, Rome). FAO Fisheries Reports, No. 82. 25 pp.

———. 1970. *Yearbook of Fishery Statistics, 1969.* Vol. 28.

Fujino, K. 1969. "Atlantic Skipjack Tuna Genetically Distinct from Pacific Specimens." *Copeia,* No. 3, pp. 626–29.

Gulland, J. A., and J. E. Carroz. 1968. "Management of Fishery Resources." In S. F. Russel and M. Younge (eds.), *Advances in Marine Biology,* 6:1–71. London and New York: Academic Press.

Inter-American Tropical Tuna Commission. 1971. *Annual Report of the Inter-American Tropical Tuna Commission, 1970.* English and Spanish.

Joseph, J. 1970. "Management of Tropical Tunas in the Eastern Pacific Ocean." *Transactions of the American Fisheries Society,* 99:629–48.

———, and T. P. Calkins. *Population Dynamics of the Skipjack Tuna in the Eastern Pacific Ocean* (in English and Spanish). *Bulletin of the Inter-American Tropical Tuna Commission,* Vol. 13, No. 1. 273 pp.

Kask, J. L. 1968. "Tuna: A World Resource." Unpublished. 26 pp.

Rothschild, B. J. 1965. *Hypothesis on the Origin of Exploited Skipjack Tuna* (Katsuwonus pelamis) *in the Eastern and Central Pacific Ocean.* U.S. Fish and Wildlife Service, Special Science Report on Fisheries No. 512. 20 pp.

———. 1971. *A Systems View of Fishery Management with Some Notes on Tuna Fisheries.* Food and Agriculture Organization, Fisheries Technical Paper No. 106. 33 pp.

Ryther, J. H. 1969. "Photosynthesis and Fish Production in the Sea." *Science,* 166:72–76.

Schaefer, M. B. 1968. "Economic and Social Needs for Marine Resources." In J. F. Brahtz (ed.), *Ocean Engineering,* pp. 6–37. New York: John Wiley & Sons.

7

Indian Ocean Fishery Development

JOHN C. MARR

The Indian Ocean is bounded on three sides by a large number (forty-one if island groups are included) and variety of political entities comprising one thousand million people, or roughly one third of earth's population. Table 7 has been chosen deliberately, of course, to emphasize the different character of fisheries as compared to most of the other subjects or problems commonly studied in the ocean. Fisheries do not depend upon the existence of a resource alone, but rather are the interaction of a resource and people. Their relevance is in their relation to the physical and, in a broader sense, the economic well-being of the people. Thus, while a monsoon system, or an upwelling system, or zooplankton distribution, or fish migration may all relate to people in one way or another, they also have an existence apart from their relationship to people. Fisheries do not have such an independent existence. I believe this to be a fundamental dissimilarity.

In this context, then, and since there is little point in attempting to repeat the several excellent papers which treat Indian Ocean fisheries in considerable detail (see, for example: Cushing, 1971; Panikkar, 1969; Panikkar and Dwivedi, 1966; Prasad et al., 1970; Prasad and Nair, in press; Shomura, 1972; and Shomura et al., 1967), I will examine very broadly the fisheries of the Indian Ocean, the kinds and distribution of the resources upon which they are based, the growth rates of the fisheries and their levels of exploitation, and make some brief comparison with the fisheries of the other oceans. I will then proceed to various estimates of the potential yields, how these appear on a mid-range time projection, and what their economic impact might be. I will consider some of the factors which inhibit fishery development and identify various manage-

John C. Marr is program leader of the United Nations Development Program/Food and Agriculture Organization's Indian Ocean Program. He has served as director of the Hawaii Area, United States Bureau of Commercial Fisheries, and director of what is now known as the La Jolla Laboratory of the Southwest Fisheries Center, National Marine Fisheries Service. He was a Guggenheim Fellow, assistant international co-ordinator for fisheries of the Cooperative Study of the Kuroshio and Adjacent Seas, and chairman of the Indo-Pacific Fisheries Council.

The author wishes to acknowledge the assistance of Mr. K. M. Joseph, deputy

TABLE 7

COUNTRIES BORDERING OR IN THE INDIAN OCEAN

1. South Africa	15. People's Democratic Republic of Yemen	27. Burma
2. Mozambique	16. Sultanate of Oman	28. Thailand
3. Tanzania	17. United Arab Emirates	29. Malaysia
4. Kenya	18. Qatar	30. Singapore
5. Somalia	19. Bahrein	31. Indonesia
6. French Territory A/I	20. Kuwait	32. Australia
7. Ethiopia	21. Iraq	33. Maldives
8. Sudan	22. Iran	34. Seychelles
9. UAR	23. Pakistan	35. Mauritius
10. Sinai	24. India	36. Réunion
11. Israel	25. Ceylon	37. Madagascar
12. Jordan	26. Bangladesh	38. Comoro
13. Saudi Arabia		39. Laccadives
14. Yemen Arab Republic		40. Andamans
		41. Nicobars

ment needs. I will comment on the relation between Indian Ocean fisheries and their development and the International Indian Ocean Expedition. Finally, I will describe the "Indian Ocean Program," an international attempt—an experiment if you will—to meet the multiple problems of international fishery development.

Present Status

The 1968 landings from the Indian Ocean fisheries were not quite 2.4 million metric tons or, with due allowance for growth, slightly more than 2.6 million metric tons by 1970. This is slightly less than 4.5 percent of the world marine fish catch, as can be seen in Table 8. As this table also illustrates, the Indian Ocean landings are disproportionately low compared to the surface area of the Indian Ocean and to the number of people living in the countries bordering the Indian Ocean. Even so, perhaps as many as four million people and 650,000 vessels (of all types) are involved in producing this catch.

If the Indian Ocean is divided into an eastern and western half, as it is for Food and Agriculture Organization statistical purposes, the catch of the western half is seen to be somewhat greater than is that for the eastern half. It might be thought that the catch is more or less evenly distributed around the Indian Ocean but in actual fact almost half the catch is landed in India. If we take into acount the fact that a very large

director, Offshore Fishing Station, Government of India, Cochin, in the preparation of some sections of this paper.

proportion of the Indian Ocean is located between the tropics, the general nature of the catches, as compared to the rest of the world, is about what one would expect, that is, "cods, hakes, haddocks" are replaced by "redfishes, basses, congers" and "crustacea" are more important than "molluscs" (to use FAO statistical categories), as is shown in Table 9. The largest category is "miscellaneous," which is really not so much miscellaneous as it is those landings which simply are not reported by species or groups. If the "miscellaneous" fishes are prorated between demersal and shoaling-pelagic fishes on the basis of the abundance of those groups in the reported catch, then the percentages of these fishes in the Indian Ocean catch are almost identical with their percentages in the catch of the rest of the world. There is a major difference, of course, in the great species diversity in the tropics as contrasted to the higher latitudes. In particular, the richness of the Indo-Malay fish fauna is too well known to require further comment.

TABLE 8

Comparison of Indian Ocean with All "Other" Ocean Areas
(By Percent)

	Indian Ocean	Other
Population	29.5	70.5
Fish landings	4.4	95.6
Area	20.8	79.2

TABLE 9

Comparison of the "Group" Composition of Indian Ocean Fish Landings with Landings from All Other Ocean Areas
(By Percent)

	Indian Ocean	Other
Cods, hakes, haddocks	0.2	17.6
Redfishes, basses, congers	13.1	5.4
Herrings, sardines, anchovies	19.6	37.2
Tunas, bonitas, billfishes	11.7	2.1
Mackerels, snoeks, cutlassfishes	5.2	5.5
Miscellaneous	33.0	14.4
Crustacea	7.5	2.3
Molluscs	0.6	6.3

Note: Only groups contributing 5 percent or more in either area are included.

Indian Ocean fisheries have been growing at the average annual rate of 5.95 percent over the period 1964–68, as compared with the rate of 5.43 percent for the rest of the world. (One must temper this comparison with the realization that some of the apparent increase may simply reflect an improvement in statistical reporting rather than an actual increase in landings.) The distribution of the fisheries carried out by Indian Ocean countries is primarily still coastal and primarily still restricted to relatively shallow areas. Many of the resources are still only lightly exploited or not at all, so that a substantial increase in total production is possible. In some areas, however, such as those close to the ports of Cochin, Bombay, or Karachi or along the west coast of Thailand, there are undoubtedly fisheries which either already are, or are very close to being, overexploited in the sense of being at or beyond the level of "maximum sustained yield" and of being overcapitalized at the producer's level.

The high seas fisheries of the Indian Ocean are the longline fisheries carried out for albacore, bigeye, bluefin, and yellowfin tunas, with incidental catches of billfishes, sharks, and so on. These are carried out almost entirely by non–Indian Ocean countries; specifically, Japan, Republic of China (Taiwan), and Republic of Korea. It is rather generally accepted that the tuna longline fishery for these species is either approaching, at, or beyond (depending upon the species) the level of maximum sustained yields and, in some cases at least, the fishery is already overcapitalized in terms of numbers of boats.

One interesting feature of Indian Ocean fisheries is the almost universal preoccupation of governments with shrimp and tuna fisheries. Presumably their interest exists because of the expanding international markets for the products of these fisheries and their consequent ability to earn foreign currency. The shrimp resources are largely in territorial waters, so that the shrimp fisheries are carried out domestically or by joint ventures with firms from non–Indian Ocean countries. The tuna resources, on the other hand, are largely in international waters and the fisheries for these resources are carried out by non–Indian Ocean countries, as I have already indicated. Thus, the interest of the Indian Ocean countries in the tuna fishery, at least in part, is the interest of the coastal state wishing to share in the benefits arising from an adjacent resource, even though their "adjacent" resources may be several hundreds of miles (or even in excess of a thousand miles) distant. It should be added that most of this interest in enjoying the benefits of the tuna resource relates to some future time; that is, even if the coastal state is not, for one reason or another, prepared to enter the fishery now, it understandably wishes to preserve its opportunity to do so in the future.

Potential Yields

A number of estimates of the potential yield of the Indian Ocean based on primary productivity, exploratory fishing, surface area or shelf area, and so forth, have been made. While there may be some disagreement or uncertainty as to whether the present yield can be increased by a factor of five, seven, ten, or even greater, there is general agreement that the present yield can be increased substantially with existing technology. A convenient, although perhaps conservative, estimate to use is that of Shomura (1972), who estimated the potential yield to be about fourteen million metric tons, plus "several hundreds of thousands of tons" of squid. His estimates, by major groups, are shown in Table 10.

Also shown in the table are estimates of the 1968 catch (estimates in the sense that the miscellaneous landings have been prorated between demersal and shoaling-pelagic, as previously explained), projection of the landings over a twenty-year period at a 5 percent rate of increase, and projection of the landings over a twenty-year period at an 8 percent rate of increase (Marr et al., 1971). Thus, from the standpoint of the resource base, growth rates of both 5 and 8 percent are feasible for both demersal and shoaling-pelagic fishes. In fact, the rate for the demersal group could even be substantially higher. It could not be substantially higher for the pelagic group (unless its potential has been seriously underestimated).

Clearly, however, the potential of both tuna and crustacea will be reached long before the end of the twenty-year period. Even if the potential of tuna is taken to be 500,000 metric tons and the potential of shrimp is taken to be 300,000 metric tons, these potentials will be attained in about ten years (slightly over ten years at 5 percent and slightly less than ten years at 8 percent).

Although neither I, nor, I suspect, those who have made these estimates and projections, would care to defend them in detail, I believe them to be useful approximations of relevance to those in government and industry who are concerned with fisheries and fishery development in the Indian Ocean. The uncertainties may be compounded by estimating the value of the estimated potential. This has been done in Table 11 (Marr et al., 1971) in which are shown the estimated potential increase in landings, a likely price per ton, and the value of the estimated increase, which is not quite $450 million. With value added in handling, processing, storage, transportation, and marketing, this will come to $1,793 million. If perhaps as much as one fifth of the total resources might be associated with one highly productive area, the landed value would

TABLE 10

Potential Growth Rates of Indian Ocean Fisheries
(In Millions of Metric Tons)

Category	Estimated 1968 Catch	5% per Annum Increase, 20-Year Projection	8% per Annum Increase, 20-Year Projection	Estimated Potential (Shomura, 1972)
Demersal	0.748	1.985	3.487	7.500
Shoaling-pelagic	1.145	3.038	5.336	6.000
Tunas	0.276	0.732	1.285	0.300
Crustacea	0.176	0.467	0.820	0.250

TABLE 11

Estimated Value of Indian Ocean Fisheries

Category	Estimated 1968 Catch (in millions of metric tons)	Estimated Potential Catch (in millions of metric tons)	Potential Increase (in millions of metric tons)	Estimated Price per Ton (1970 U.S. $)	Estimated Value Increased Landings (in millions of 1970 U.S. $)
Demersal	0.750	7.500	6.750	40	270.000
Shoaling-pelagic	1.150	6.000	4.850	20	97.000
Tunas	0.275	0.300	0.025	250	6.250
Crustacea	0.175	0.250	0.075	1,000	75.000
Total	2.350	14.050	11.700		448.250

Value added: 448.250 × 4 = 1,793.00 in millions of US $.
Production area (1/5 of total): landed value 89.650 million of 1970 U.S. $.
Value added 358.600 millions of 1970 U.S. $.

be $89,650,000 and the value added $358,600,000. Although these are gross rather than net values, it is apparent that fishery development in the Indian Ocean can make a significant contribution to general economic development.

Inhibiting Factors

With these valuable potentials more or less close at hand, one may wonder why they have not been brought into production at a faster rate. There are, of course, a variety of reasons why this is so. One of these, often overlooked, is simply timeliness. The large *Sardinella* population of the northwestern Arabian Sea is now being investigated as a likely source of additional supplies of fish meal. In 1950, or 1955, or even more recently, this would not have been a timely enterprise, since the world market for fish meal had not attained its present volume and there were then other sources of supply adequate to meet the demand. A similar situation exists with respect to the skipjack tuna resource. It is really only now that the world demand and supply situation is such that it is timely to harvest the Indian Ocean skipjack for the world market.

Another factor which has to some extent inhibited fishery development in the Indian Ocean has been lack of resource knowledge. However, knowledge of this area has been increased greatly in recent years through the work of national projects, bi- and multilateral projects, by private companies, and, to some extent, by the International Indian Ocean Expedition (IIOE).

Lack of capital, technology, and managerial skills are almost standard inhibitors of development in any field. They obtain in varying degree, depending upon locality, with respect to Indian Ocean fishery development. These needs may sometimes be met by joint venture operations in which the Indian Ocean country could supply proximity to the resource and access to labor supplies, while an economically developed country outside the Indian Ocean could supply the capital, technology, and managerial skills. And, of course, there are examples of such joint ventures now operating in the Indian Ocean region. Unfortunately, there are frequently sources of friction, misunderstanding, and suspicion in joint venture operations and I do not believe that these operations have been used to anywhere near their full potential. As Myint (1970:58) has put it: ". . . the southeast Asian countries can expect to enjoy a rapid rate of economic development through export expansion in the 1970's, provided that an effective link can be forged to connect their abundant natural resources with the expanding world-market demand for their exports . . . the effectiveness of this link . . . will depend on the policies to promote freer entry of private foreign investment. . . ."

Other factors tending to inhibit fishery development in the Indian Ocean include the lack of infrastructure and other facilities, institutional barriers, communications, and cultural characteristics. Among the Indian Ocean countries there is wide variation in the infrastructure development—in the existence, for example, of harbors and road or other transportation systems, and other facilities, such as commercial banks. In some areas these are well developed and in others they are virtually lacking. Institutional barriers are many and varied and include such things as seemingly conflicting government policies; for example, (1) an import restriction policy which makes it difficult to get spare parts for engines or diesel fuel, as a result of which fishing effort is restricted, (2) excessive requirements for the licensing of fishing vessel masters, with a resulting shortage of masters and restriction of fishing effort, or (3) overvaluation of currency which, among other things, inhibits the development of export industries. Communications, or rather the lack of them, between government and industry or even within government inhibits fishery development through the nontransmittal of government exploratory fishing results to industry, government unawareness of industry problems, and so on. Finally, in some Indian Ocean countries fishermen have been at the low end of the spectrum of social acceptance and fishing is therefore not a career to which it is easy to attract young people.

Management Needs

Management needs include, on a day-to-day basis, managerial skills and the necessary information for rational decision making. On a longer range basis, there are also needed solutions to some rather general, fundamental problems. Managerial skills are of those kinds which can be formally learned (e.g., accounting, application of technological advances, etc.) and those which represent a way of thinking (e.g., ability to question, to innovate). Four examples, drawn from Tussing (1971), illustrate the management problems in the region. First, the optimum capacity for a processing plant with a fluctuating supply of raw material both within and between seasons is generally considered to be that capable of handling the peak supply. If, however, the frequency of peaks, the difference between the peaks and the troughs or peaks and the mean, and the profitability per unit are taken into account the conclusions might be quite different. Second, expensive equipment is used at a fraction of capacity, even in the face of unfulfilled demands for its services, because it is used on only one shift per day. Third, capital is wasted by over- or underprovision of capacity (e.g., one plant had three stand-by diesel generating units to provide electricity during periods in which the local utility service was impaired, when one, or at most two, would have been adequate). Finally, capital is also wasted by the use of labor-consuming methods (e.g., doors on plate freezers are left open for twenty to thirty minutes on each load through the lack of an inexpensive modular loading device).

In addition to the kind of information about internal operations requisite for rational decision making, there is also needed information, for example, about the status of the resource upon which an industry is based. (And this kind of information is obviously required by managers in both industry and government.) For a variety of reasons, the collection and dissemination of such information falls upon government. It is seldom easy, if even possible, to obtain in ideal form and less than perfect, innovative approaches to obtaining such information must be developed and used. Through the use of such information it is possible to prevent, or minimize, biological and economic overfishing.

There is a group of fundamental problems, solutions of which will be of importance in the Indian Ocean, as well as elsewhere in the world. These problems have to do with such things as (1) stock and recruitment, (2) variations in the distribution and abundance of a resource in relation to variations in the environment, (3) ecological effects of selective or nonselective harvesting, and (4) conceptual models in the area of population dynamics–resource management. I mention these here, not only to call attention to the problems themselves, but also to call attention to the need to attack them by problem-oriented studies rather than as part of species-oriented studies. These kinds of problems can be attacked anywhere in the world. They are particularly appropriate for university laboratories.

Indian Ocean Fisheries and the IIOE

It is, I think, reasonable to inquire as to the impact of the IIOE on fishery development in the Indian Ocean, for, as Panikkar (1966) has said, "When the programs of the International Indian Ocean Expedition were formulated, fisheries potential was used as an impressive argument to stimulate interest in the project in the Asian and African countries." And, I would add, in other parts of the world as well. Further, ". . . the actual fisheries work accomplished during the expedition itself has been disappointingly small. . . ." This is borne out by examination of the IIOE Collected Reprints (UNESCO, 1965–69). Of the total of almost five hundred papers, only ten dealt with fisheries and only another twenty-three dealt with organisms that were or conceivably might be the objects of fisheries even though the papers themselves were not fishery-oriented. Thus, there may be some disillusion with the IIOE, particularly within the Indian Ocean region.

Still, there are benefits to fishery development from the IIOE. These may be categorized as follows, more or less in increasing order of direct relevance to fisheries:

1. The general store of information about the Indian Ocean has been greatly increased. Some of this information (in addition to that listed in item 4 below) will eventually prove to be of relevance to fishery development, most likely in completely unexpected ways.

2. There is a continuing interest in the Indian Ocean on the part of the world marine science community, as a result of which additions to knowledge about the Indian Ocean will continue to accrue.

3. There is a heightened interest in marine science on the part of some, if not many, of the Indian Ocean countries, which will also result in additions to knowledge about the Indian Ocean.

4. There is a body of information on such features, for example, as the distribution of the upwelling–high productivity areas, the depth distribution of the oxygen minimum, and the distribution and abundance of fish larvae, all of which have rather obvious relationships to fishery development.

It has been the philosophy of some that any increase in man's knowledge and understanding of the ocean in general is bound to also increase man's knowledge and understanding of fishery resources in particular and, hence, to facilitate fishery development. In this general context, there is no question but that the IIOE has contributed greatly, as I have indicated. In the specific context of fishery development, however, and with the benefit of hindsight, I am left with the reservation that a much greater impact could have been achieved much more efficiently.

Indian Ocean Program

I would like to turn now to the Indian Ocean Program. This program, or the International Indian Ocean Fishery Survey and Development Program to use its full name, undoubtedly owes its origin to three circumstances: (1) the large population around the Indian Ocean, (2) the three million metric tons annual protein deficit in the diet of this population, and (3) the large potential for increased yields from Indian Ocean fishery resources. Wib Chapman was one of the first to appreciate the potential of the Indian Ocean and the opportunity it presented for regional development. With his persuasiveness and mobility, he effectively presented this view in various FAO fora and elsewhere. If Chapman can be considered the informal originator of the Indian Ocean

Program, the formal sponsor is the Indian Ocean Fishery Commission. At its first session in September 1968 the commission asked FAO to arrange with the United Nations Development Program for a planning phase of the program. This was done and I became involved early in 1970.

Most of 1970 was taken up with attempting to become familiar with the region—its places, people, problems, and possibilities. A series of summary-review reports were prepared by consultants on resource inventory; distribution and abundance of fish eggs and larvae; stock assessment; management; statistics; survey and charting of resources; environmental research; experimental fishing; vessels and equipment; shipyards; harbors; handling and processing facilities; economic characteristics in the development process; fishery economics; international trade in tuna, fish meal, shrimp, crabs, and groundfish; and economic planning for fishery development. Most of these have now been published.

Using these and other sources of information, Ghosh, Pontecorvo, Rothschild, Tussing, and I prepared in January 1971 *A Plan for Fishery Development in the Indian Ocean.* This plan contains a large number of specific proposals which need not be described here. It is sufficient to say that they are proposals for: (1) investment projects, (2) preinvestment resource assessment surveys, (3) feasibility and other studies, (4) management and institutional arrangements, and (5) training and conferences.

This plan was approved in principle for the Indian Ocean Fishery Commission by its executive committee in the latter part of April and has now been distributed to IOFC member governments, other interested governments, the UNDP, and world and regional banks.

The recommended proposals will be carried out under UNDP or other multilateral funding, through bilateral programs, and by private investment. I have spent the remainder of 1971 facilitating such arrangements, and a full-scale program in 1973–76 is anticipated. The UNDP has approved a modest operational program for 1972. It is not likely that any large-scale, new operational activities could get underway before the second half of 1973.

I would like to mention in passing—with the vested interest of blood, sweat, and tears—that I believe this is the first attempt to examine fishery development on a broad regional basis and it is also, I believe, the first time in which fishery development has been examined in a broad matrix of economic development on a regional basis.

Let me finish with two topics in the context of this series of papers: (1) the objective of the Indian Ocean Program and (2) some of the problems to be met in achieving this objective.

The objective of Indian Ocean fishery development is to contribute to general economic development leading to generally increasing standards of living and to the diffusion of economic well-being throughout the population. As a quantifying tool we may find it convenient or necessary to talk about GNP or rate of growth of GNP or GNP/capita, but we should not forget that what we are really talking about are such basic things as more food, shelter from the elements, access to medical attention, the knowledge that your children may be literate even if you are not, a job.

Now as to the problems: First of all, one cannot get very far into development in the Indian Ocean region without being overwhelmed by a pressing sense of urgency. One has only to recall the recent news of events in the Indian Ocean region to realize that time is of the essence. What this means is that those who are involved in development activities have to take decisions on the basis of far from complete information. Indeed, because of this urgency it is essential to develop rapid, innovative ways of looking at old problems.

The second and third problems are interrelated and might well be taken up together. How can the fishery resources of the Indian Ocean be developed for the benefit of the Indian Ocean countries? It doesn't make much sense to go through all the developmental work and expense if this will benefit only countries outside the Indian Ocean simply because they are better able to take advantage of the situation. On the other hand, common property resources in the Indian Ocean are just as available to non–Indian Ocean countries as to Indian Ocean countries. The resolution, it seems to me, is to bring together the access to resources and labor supplies of the Indian Ocean countries with the technology and capital of non–Indian Ocean countries in a way that is mutually beneficial. There needs to be constructive attention to improved joint venture arrangements, which minimize the opportunity for individuals on either side of the joint venture to put their personal interests first.

A fourth problem is that of guiding development so as to avoid serious overcapitalization in countries where capital is scarce. The situation in most Indian Ocean countries is such as to suggest that this possibility has fewer obstacles, or at least a different set, than would, say, obtain with respect to the fisheries of the Pacific Northwest.

The fifth and last problem I will mention is that of the management of common-use resources. Here I make no brief that our approach will be more effective than other approaches to this problem elsewhere in the world. I can only say that we may be in a position to anticipate the problems and we already have a forum, the IOFC, in which they can be met.

May 1971

REFERENCES

Cushing, D. H. 1971. *Survey of Resources in the Indian Ocean and Indonesian Area.* Food and Agriculture Organization, Indian Ocean Fishery Commission, IOFC/DEV/71/No. 2. Rome: FAO. 123 pp.

Marr, J. C., D. K. Ghosh, G. Pontecorvo, B. J. Rothschild, and A. J. Tussing. 1971. *A Plan for Fishery Development in the Indian Ocean.* Food and Agriculture Organization, Indian Ocean Fishery Commission, IOFC/DEV/71/ No. 1. Rome: FAO. 78 pp.

Myint, H. 1970. *Southeast Asia's Economy in the 1970's.* Chapter 1: "Overall Report, Asian Development Bank." Manila: Asian Development Bank.

Panikkar, N. K. 1966. "Fishery Resources of the Indian Ocean." *Current Science,* 35 (No. 18): 451–55.

———. 1969. "Fishery Resources of the Indian Ocean." *Proceedings of the Symposium on the Indian Ocean,* pp. 811–32. Bulletin of the National Institute of Sciences of India No. 38, Pt. 2.

———, and S. N. Dwivedi. 1966. "Fisheries of the Asian Countries Bordering the Indian Ocean." In *Proceedings of the Seventh International Congress of Nutrition,* Vol. 4.

Prasad, R. Raghu, S. K. Banerji, and P. V. Ramachandran Nair. 1970. "A Quantitative Assessment of the Potential Fishery Resources of the Indian Ocean and Adjoining Seas." *Indian Journal of Animal Science,* 40 (No. 1): 73–98.

Prasad, R. Raghu, and P. V. Ramachandran Nair. In press. "India and the Indian Ocean Fisheries." In *Proceedings of the Symposium on the Indian Ocean and Adjacent Seas.* Marine Biological Association of India.

Shomura, R. S. 1972. "Indian Ocean Coastal Waters." In J. A. Gulland, *The Fish Resources of the Ocean.* London: Fishing News (Books), Ltd.

———, D. Menasveta, A. Suda, and F. Talbot. 1967. *The Present Status of Fisheries and Assessment of Potential Resources of the Indian Ocean and Adjacent Seas.* Food and Agriculture Organization Fisheries Reports, No. 54. 32 pp.

Tussing, A. R. 1971. *Economic Planning for Fishery Development.* Food and Agriculture Organization, Indian Ocean Fishery Commission, IOFC/DEV/ 71/No. 19. Rome: FAO. 26 pp.

United Nations, Educational, Scientific, and Cultural Organization. 1965–69. *International Indian Ocean Expedition.* Collected Reprints, Vol. I–VI.

8

Jeffersonian Democracy and the Fisheries

J. L. McHUGH

The great American dream of Thomas Jefferson and his fellow founders of the American republic was a government of the people. Jefferson himself stated it this way (Dewey, 1940):

> Were not this great country already divided into States, that division must be made, that each might do for itself what concerns itself directly, and what it can so much better do than a distant authority. Every State again is divided into counties, each to take care of what lies within its local bounds; each county again into townships or wards, to manage minuter details; and every ward into farms, to be governed each by its individual proprietor.

By this method of government, said Jefferson: ". . . every man in the State would thus become an acting member of the common government."

This concept was reiterated by President Richard M. Nixon in his 1971 State of the Union message:

> The farther away government is from people, the stronger government becomes and the weaker people become. Local government is the government closest to the people and more responsive to the individual person; it is people's government in a far more intimate way than the government in Washington can ever be.

And to turn rhetoric into action he proposed to return some of the federal revenues to the states and local governments.

The Jeffersonian form of government has deteriorated steadily since the Constitution entered into force. Even where the form of true republi-

J. L. McHugh is presently professor of marine resources at the Marine Sciences Research Center, State University of New York, Stony Brook. He has been head of the Office for the International Decade of Ocean Exploration, National Science Foundation; acting director, Office of Marine Resources, Department of the Interior; chief, then assistant director, Division of Biological Research, Bureau of Commercial Fisheries, and finally deputy director of the bureau; and director of the Virginia Fisheries Laboratory. He has served in numerous international commissions and has

can government has been retained, it is not true people's government. As Arthur Schlesinger, Jr. (1971), has pointed out: ". . . local government is characteristically the government of the locally powerful, not of the locally powerless; and the best way the locally powerless have found to sustain their rights against the locally powerful is through resort to the national government."

Fisheries in the United States have adhered rather rigidly to the form of Jeffersonianism, if not entirely to its practices. Fishery laws and regulations and their enforcement have remained the responsibility of the individual states, and in some cases the counties or even the towns. Most states have separate administrations for freshwater fisheries, which are mostly recreational, and for marine fisheries, which are mostly commercial in total weight landed but have important and growing recreational elements. Some states even have a distinct and separate shellfishery commission or shellfish agency as well. In some states the tidewater counties may make their own fishery regulations, especially for shellfisheries, and in some parts of the northeastern United States clam harvesting in the littoral zone is regulated by the individual towns.

It is granted that there may be compelling social as well as economic or scientific reasons for local control of fishing. It is especially logical that landlocked resources like most freshwater fish stocks be subject to local control. In many respects, also, fisheries for sedentary species like oysters and clams can be controlled best by local authorities, although it will be shown later that under some circumstances local control has destroyed, rather than enhanced, some important shellfisheries. With respect to migratory species, including nearly all marine fishes and some important invertebrates that move freely through the waters of more than one state, local control is anything but logical or beneficial, either to man or to the resource, yet local control has persisted in the United States for almost two hundred years, and this privilege is defended just as stoutly now as it was when regulated commercial fishing first began in United States waters.

The east coast of the United States offers some especially interesting examples of the irrationality of local control. I will use Chesapeake Bay as a source of such case histories because this is the area with which I am most familiar at first hand. Chesapeake Bay provides good examples also because its waters and fishery resources are nearly equally divided between two states, Maryland and Virginia. The oyster industry has always dominated the fisheries of these two states, and it produces a harvest with a landed value greater than the value of all its other fishery resources. Oystering is so traditionally dominant in Virginia

---

represented the United States on the Inter-American Tropical Tuna Commission and the International Whaling Commission.

that the inspectors responsible for fishery surveillance and law enforcement are locally referred to as oyster inspectors, not fishery inspectors.

The Chesapeake Oyster Industry

A brief historical review is necessary to understand the complicated structure of the oyster industry in the Chesapeake area.

Public Grounds

Late in the nineteenth century in Virginia and early in the twentieth century in Maryland it was decided that the natural oyster grounds were public property, open to harvesting by any citizen of that state who chose to pay a modest license fee and who agreed to abide by the oystering laws and regulations. These grounds were surveyed by the Coast and Geodetic Survey and large-scale charts of these public grounds were prepared. Like all such surveys, these were not entirely accurate. Some barren areas, unsuitable for natural oyster production, were included inadvertently, and some naturally productive grounds were omitted. But the errors were relatively minor, and these surveys, named the Baylor Survey in Virginia and the Yates Survey in Maryland, after the officers of the Coast and Geodetic Survey in charge of the two projects, still form the basis today of the public oyster management programs of the two states.

Each state has a substantial management program which specifies the kinds of harvesting gear which can be used on the public grounds and prescribes harvesting practices. Each state also conducts extensive rehabilitation activities, such as shell planting to provide a base of attachment for oyster larvae when they are ready to set. In Virginia oysters can be taken from the public grounds only by the laborious method of "tonging." On most public grounds only hand tongs are allowed. Essentially these tongs are a pair of rakes connected to baskets made of iron rods. The handles of the rakes, up to twenty-six feet long, are pinioned near the lower end. The oysterman stands on the gunwale of his small boat and gropes for oysters with his heavy pincerlike device, and when he feels that he has a load he hauls the cumbersome contraption to the deck. The work requires strength and stamina. On certain public grounds "patent tongs" may be used. These are similar, but with short, metal handles. They are lowered by a cable which runs through a block on the boom of the vessel with the jaws locked open. When the gear hits bottom a tripping device is actuated, the jaws close, and the gear is then hauled to the surface by power, and the load dumped on deck.

In Maryland, tongs also are permitted on the public grounds, but dredging, which is prohibited on public grounds in Virginia, also is allowed under some circumstances. Until recently the vessels which pulled the dredges could not use motor power, only sails. Today this restriction has been relaxed partially, so that power is permitted on Monday and Tuesday during the oystering season, sail the rest of the week. These restrictions on harvesting efficiency were imposed as conservation measures but they have no evident conservation value in themselves. Instead, they serve to distribute the catch among a greater number of people.

In both states, except on grounds designated as seed oyster areas, a minimum size limit is imposed, usually three inches. The catch must be "culled" to return undersized oysters and dead shells back to the bottom. In seed oyster areas, such as the famous James River seed area in Virginia, small oysters may be taken for replanting on private grounds. A few market-sized oysters may appear in the catch from seed areas, but these are only a minor part of the total harvest. Culling also is required on the seed grounds, to conserve the supply of shell for cultch.

Virginia requires that oyster shucking houses reserve a certain proportion of their empty shell for purchase by the state for replanting on the public grounds. Maryland does so also, but her management program is more extensive than Virginia's, as might be expected in an industry based primarily on public management (Manning, 1969). Maryland also transplants living oysters from one ground to another as required to provide brood stocks for production of larvae or to stimulate maximum growth.

Private Grounds

Most of the bottom in Chesapeake Bay, in both states, is unsuitable for natural oyster production. The principal reason is that the bottom is too soft to support shells or objects with firm surfaces suitable for setting of spat. Occasionally, a large object may lodge on the bottom and fail to sink entirely into the mud. If the surface is suitable for setting, and larvae are available, a small colony of oysters may become established, and may spread slowly, as oysters grow and die and a mass of shell accumulates. But this usually is apt to be a transitory phenomenon. Some barren bottom can be made suitable for oyster production by planting shell or other material, and in the bay as a whole some 150,000 acres have been leased by the states to private oyster planters, who prepare the bottom and then plant seed oysters. If conditions are favorable, some of these grounds may even become self-sustaining.

Although both states lease bottoms outside the public grounds for private planting, they differ widely in their attitudes and policies with respect to leasing. Virginia traditionally has encouraged private enterprise, and about 140,000 acres (roughly 225 square miles) of bottom is presently leased by private planters. Most of Virginia's market oyster production comes from these grounds. Maryland, on the other hand, traditionally has supported a policy of public production, and she currently has slightly more than 10,000 acres under lease. In Maryland, barren bottom can be leased to individuals but not to corporations and maximum lease is thirty acres. Less than 10 percent of the annual harvest of market oysters comes from private grounds.

Public vs. Private Management

Virginia has long justified her liberal policy of private oyster planting by pointing out that most of her market oyster production comes from private grounds, and that total oyster production in the state has held up better than in Maryland. The argument based on private production is somewhat misleading, however, because most of the oysters produced by private planters came originally from public grounds, mainly in the James River. Until recently, private production has been responsible for maintaining the Virginia harvest of market oysters, but in the last few years Maryland has intensified her management program on the public grounds, and by relatively large expenditures has shown that public management can be successful in improving yields. Thus, oyster production in Maryland has been rising since 1963 while production in Virginia has been dealt a severe blow by epizootics such as the fungus *Dermocystidium marinum* and more recently by the haplosporidian once known as MSX, now identified as *Minchinia nelsoni.*

The historical record, however, demonstrates very clearly that oyster management in Chesapeake Bay has failed, if maintenance of historic yields is a valid criterion. Total production has fallen rather steadily from about 117 million pounds of meats in 1880 to less than 25 million in 1970, a reduction of about 80 percent in ninety years. Total oyster production along the Atlantic coast has fallen proportionately. Some of this drop may be indicative of changing demand, but apparently not all of it, for Chesapeake Bay, once the leading producer of oysters in the United States, yielding nearly 80 percent of total domestic landings, now produces only about 40 percent. Moreover, domestic production of oysters now supplies only about 80 percent of total domestic consumption, whereas in 1920 imports were negligible. By any standard the Chesapeake Bay states have not managed their oyster resources well, nor have the private planters.

With respect to the present program in Maryland the statements above might be questioned. From 1963 to 1970 oyster production in Maryland has risen from about 7.75 million pounds to about 15 million pounds of meats, an increase of about 7 million pounds with a landed value of about nine million dollars. The traditional opposition in Maryland to private planting is apparently less intense today. But the increase in production has been achieved by an intensive public planting program at considerable cost. A benefit-cost analysis would be useful to measure the net effect of this program.

Scientific Research on Oysters

Someone has said that the oyster has been studied by scientists in more detail than any other marine animal. Whether that is true or not, there is no question that a great deal has been learned about this important commercial animal. Korringa (1952) summarized the status of knowledge up to 1951 in an important paper that will remain a most useful reference. Yonge's (1960) book and the more recent monograph by Galtsoff (1964) complement Korringa's work and show clearly how much has been learned in the short span of about two hundred years. The sad thing is that so little of this knowledge has been put to use by Chesapeake Bay oystermen. It is informative to review some of the major advances in knowledge of oyster biology in Chesapeake Bay, to consider the potential value of this information, and to determine how well this potential has been exploited. Actually, this research has not told us very much that a smart and aggressive oysterman could not have found out for himself (Andrews and Hewatt, 1951). Yet, with few exceptions, oystermen have not taken the initiative for themselves, nor have they been eager to try the suggestions of scientists. As will be demonstrated later they have been able to make a profit using the crude farming techniques of the past, and have been reluctant to change. The attitude has been one of "what was good enough for my daddy, and for my granddaddy before him, is good enough for me." This conservative attitude has been maintained and defended in the face of steadily declining total yields.

Some outstanding original scientific research has been done on oysters in Chesapeake Bay, largely by biologists of the Virginia Institute of Marine Science (formerly the Virginia Fisheries Laboratory), the Chesapeake Biological Laboratory in Maryland, and the Shellfish Laboratory of the National Marine Fisheries Service at Oxford, Maryland. Much of this research has produced results that could be of practical value to oystermen.

Ecological niches. One of the facts most obvious to a student of Chesapeake Bay oysters is that oysters do not thrive naturally in all parts of the bay. The natural oyster grounds in Virginia are for the most part in bays or estuaries, and not in the Chesapeake Bay itself. In Maryland the natural grounds are about as likely to be in the bay as in the lower parts of the rivers. It does not take much scientific detective work to conclude that the major factor controlling distribution of natural oyster grounds in the bay is salinity. Provided that the bottom is suitable for oyster production, almost all of the natural oyster grounds lie within the salinity range 7‰ to 18‰. The reason is that oysters can withstand these reduced salinities, but their principal predators and diseases cannot. This should provide lesson number one for the oysterman: if possible, select suitable ground within these salinity limits for leasing.

In salinities higher than 18‰ oysters can survive under natural conditions also, but they do so by a totally different response. This phenomenon is demonstrated most clearly in the bays and lagoons on the seaside of the eastern shore, which is the name commonly used for the eastern coast of the peninsula that forms the eastern boundary of Chesapeake Bay. Here oysters can survive only in the intertidal zone, where they evade their predators and diseases by the same general technique that is so effective in low-salinity waters. The oyster is able to withstand exposure to air at low tide, whereas its enemies cannot. The same phenomenon is exhibited within Chesapeake Bay in higher salinities, where in some areas oysters have become established in the intertidal zone, either on shore, or on pilings of piers, stakes marking boundaries of oyster grounds, navigational buoys, or other suitable habitats.

Disease. Mass mortalities are not uncommon on oyster grounds. Often these may be caused by epizootics, although this is a fairly recent discovery. The earliest and most famous case is the Malpeque disease in Prince Edward Island, Canada, where the industry was almost wiped out in the period 1915–36 (Logie, 1956). Since that time biologists have had considerable success in developing disease-resistant strains.

Many historic mass mortalities of oysters were never explained. All were not caused by disease, but it is clear that some probably were. Oyster research in the Gulf of Mexico in the 1940s and early 1950s, much of it sponsored by oil companies which were being blamed for all the ills of the oyster industry there, discovered a fungus (*Dermocystidium marinum*) which caused heavy mortality of oysters in summer (Mackin, Owen, and Collier, 1950). In the 1950s it was discovered that this fungus was present in Chesapeake Bay, principally in higher-salinity waters in Virginia (Andrews, 1955). In the mid-1950s this fungus caused heavy mortality in oysters suspended in trays from the pier of the Virginia Fisheries Laboratory. In some years as many as 25 to 50 percent of the adult oysters alive in June were dead by September or October (Andrews and Hewatt, 1957).

In the late 1950s a serious mortality of oysters began in Delaware Bay, and in a very few years this phenomenon virtually wiped out the industry in that area (Haskin, Canzonier, and Myhre, 1965). The cause was diagnosed as a disease. The organism was found, but not identified, and for several years it was given the tentative name MSX, for multinucleated sphere of unknown origin. In the early 1960s this organism reached epizootic proportions in Chesapeake Bay, with equally disastrous effects upon the oyster industry, especially in the Virginia waters of the bay itself. Several of the old family oyster planting businesses of Virginia were wiped out by this epizootic. The organism was identified as a haplospordian protozoan and it was given the name *Minchinia nelsoni* (Haskin, Stauber, and Mackin, 1966). The specific name honored the great Dr. Thurlow Nelson, once professor of biology at Rutgers University.

It has been generally agreed that mass mortalities from disease have been stimulated by careless oystering practices, such as uncontrolled transfers of oysters from one locality to another, and crowding of oysters on planted beds, perhaps favored by man-made physical and chemical changes in the environment (Sindermann and Rosenfield, 1967). The effects of known diseases, like those of predators, can be reduced to a minimum by taking advantage of their relations to temperature, salinity, and other environmental variables.

Growth and mortality. Oystermen in Chesapeake Bay have traditionally selected the time to harvest their crops by examining the size of the oysters on their beds. The size and size range can be determined easily by sampling the ground with dredges. This practice probably was adopted because larger oysters, locally known as selects and counts, bring higher prices. Size as a criterion for harvesting, however, does not provide enough information to judge when to harvest for maximum yield of oysters in the shell (Hopkins and Menzel, 1952; McHugh and Andrews, 1955). The crude method of sampling used by oystermen does not permit quantitative estimates of mortality. Yet mortality plays a very important role in determining yields, especially when it operates only at certain restricted times of the year, as *Dermocystidium* does. Investigations in Virginia in the 1950s, for example, showed clearly that oysters were being harvested from private grounds after the point of maximum biomass of oysters had been reached, and deaths had removed more biomass than was being replaced by growth (McHugh and Andrews, 1955). At that time *Dermocystidium marinum* was epizootic in the saltier waters of Chesapeake Bay, causing mortalities as high as 50 percent or higher in some areas in a period of three or four months in summer. In other words, by October about half of the oysters that were alive in May or June had died.

About the time that we had reached these conclusions and were attempting to persuade oystermen to take advantage of the knowledge gained, the Campbell Soup Company was installing equipment to produce oyster meats for its frozen oyster soup. The requirements were for oysters of a certain uniform size, and it happened that the size required was smaller than what was normally harvested. In meeting these requirements the oyster planters concerned had to harvest their oysters sooner than usual and to their delight discovered that they reaped an unexpected dividend in the form of higher yields per unit volume of seed planted. Thus, circumstance demonstrated the validity of the scientific findings. If this fortunate concurrence of events had not come about, it is questionable whether the scientists would have been able to persuade oystermen to try this experiment for themselves. Indeed, as will be described later, there were good economic reasons why oystermen would be reluctant to harvest earlier in the year.

Annual cycle of oyster condition. The condition or quality of oysters is an important index of the yield in meats to be expected from a given quantity of oysters in the shell. Scientists derive an index of quality or "fatness" from the ratio of the volume or weight of the meat to the volume of the shell cavity. In most areas of Chesapeake Bay and its tributaries the highest index of condition comes in May or June and the lowest in August and September or later (Haven, 1962). The traditional time for harvesting oysters, when demand and prices are high, begins in October, stays high until Christmas time, and then falls off fairly steadily until summer. The myth of the "R" months still governs the demand for oysters. Some oysters are harvested in every month of the year, but in the state of New York, for example, in the decade 1960 to 1969 inclusive, less than 6 percent of the value of the annual harvest was taken in the four months lacking an "R." In contrast, about 60 percent of the year's landed value was harvested in the four-month period October to January inclusive. Thus, the major harvest begins in October, at a time when the oysters have gone through a period of stress associated with high water temperatures, little or no growth, and the physiological changes associated with spawning. Meats are thin and watery, flavor is poor, and the yield of meats per unit volume of live oysters is at a minimum. In spring and early summer oysters are storing glycogen preparatory to spawning. They are fat and flavorful, and the yield of meats may be double the yield in fall, or more. Yet demand and price are at a minimum, and the industry traditionally almost ceases to produce.

Bottom vs. off-bottom culture. In Chesapeake Bay, as in almost all of the domestic oyster industry, the crop is planted and grown on the bottom. It is well known that oysters grow better, suffer less predation, and are fatter and of more desirable shape if they are grown off the bottom on racks, in suspended trays, or on hanging ropes (Engle, 1970). Such practices have been developed to a high degree of efficiency in Japan, Australia, and other parts of the world, but oystermen in Chesapeake Bay have not adopted or even experimented seriously with such methods because they have concluded subjectively that labor costs would be prohibitive. A successful commercial oystering enterprise in trays raised off the bottom on racks was conducted in the York River, Virginia, by the Chesapeake Corporation in the late 1930s (Evans, 1943). This produced well-shaped oysters which sold for premium prices, but commercial oyster production was not the primary objective of the experiment and it was not continued.

More and more interest is being shown in off-bottom culture of seed or market-size oysters. The advantages are many, and it is likely that the more uniform production, increased yield, and higher price may compensate for the higher costs of production. Another advantage of suspending oysters is that the available food in the entire water column can be utilized, and this becomes more attractive as the area available to oystermen shrinks.

Artificial culture. Development of reliable methods of spawning oysters and rearing larvae and spat in the laboratory (Loosanoff and Davis, 1963) promises to solve one of the most frustrating problems of the oyster industry—a dependable supply of seed. Combining these techniques with controlled or semicontrolled rearing from young to market size will bring about a true oyster farming industry. Problems of seed supply and efforts to do something about it are not new. As early as the 1880s the State of New York was experimenting with artificial oyster culture (Wells, 1922) and had established a shellfish hatchery at Cold Spring Harbor on Long Island. Even earlier (1879) Professor Brooks at Johns Hopkins University had fertilized oyster eggs in the laboratory. But none of the early workers was able to bring the larvae to the setting stage. This was accomplished first in 1920, and a hatchery was established again on Long Island. This oyster hatchery program proceeded with vigor through 1926, but apparently was terminated rather abruptly shortly thereafter.

Today on Chesapeake Bay, on Long Island, and elsewhere in the United States, several commercial firms have established shellfish hatcheries. None of these is yet producing seed or market oysters on a commercial scale, but it seems just a matter of time before the process becomes routine and predictable. The more promising of the present experiments should be supported until they succeed.

Attitudes of Planters, Tongers, and Administrators

In the Chesapeake Bay region the social-political structure and the philosophies of the major segments of the seafood industry and government could scarcely be organized in a way that would be less favorable to a healthy oyster industry. Virginia, in which much of the bottom available for leasing to private planters is marginal for growing oysters, has a liberal policy toward private control of grounds that do not produce oysters naturally, but most of these grounds are marginal for oyster production. Maryland, which has extensive areas of bottom which do not produce oysters naturally, many of which are free from the major sources of mortality that make leased bottom in Virginia marginal, does not encourage private planting, or at least not until recently. Of the approximately 150,000 acres of private oyster ground in Chesapeake Bay, considerably less than 10 percent is in Maryland. Virginia, which has the most extensive and most reliable natural seed beds, prohibits exportation of seed to Maryland or other states. Extensive grounds in Maryland having a potential to produce three to five bushels of market oysters from every bushel of seed planted have been unavailable for oyster production. Even today, when legal obstacles to private leasing in Maryland have been largely abolished, there has been no rush to take advantage of the opportunity, principally because capital is not available. Most of the private grounds in Virginia yield from one half to one bushel for each bushel of seed. The oyster industry is prevented by public policy and state laws, and recently by the reluctance of bankers to make loans to planters, from taking advantage of the natural ecological characteristics of the bay that would suit it so well to scientific oyster farming.

It will have been obvious also that the traditional seasonal patterns of preference for oysters have worked against the economics of the industry. Yields of two to four or more times present yields could be obtained from the same quantity of seed if oysters were harvested in May or June instead of late fall or winter. This also would allow more crops per unit time. The Campbell Soup Company enterprise already mentioned provided a means for taking advantage of this knowledge, but this helped only some planters, because the demand for this product is not unlimited. Incidentally, this enterprise solved another difficult problem for the industry, the labor-intensive and increasingly costly operation of shucking, for which no satisfactory mechanical method has yet been developed. For soup, fresh oysters are not necessary, and meats can be removed from the shells easily and cheaply by steaming and tumbling.

Thus, it is clear that existing scientific and technical knowledge is not being used effectively by the Chesapeake Bay oyster industry, and this is true for most segments of the United States oyster business. Yet there was at least one planter in Maryland who was able to avoid the major constraints. He was one of the lucky few in the state who had rights to lease bottom for private planting. He had the foresight to go into partnership with a Virginian who had leasing rights to ground in the James River seed area outside the public grounds but equally productive of seed. It is perfectly legal to transplant seed raised on private ground in Virginia to leased ground in Maryland. The balance sheet of this tidy operation would be most interesting to see.

The principal reason for the network of institutional barriers which prevent use of Chesapeake Bay as a gigantic oyster farm is the age-old battle between the oyster tonger and the planter, or the oystermen of one area against the oystermen of another. The small entrepreneur who works the public grounds with backbreaking hand labor does not welcome the development of large-scale private planting industry, for he sees it as a threat to his "independence." He is equally resistant to oystermen from other areas within his own state who may wish to harvest his traditional grounds. The usual procedure is to persuade the state legislature to pass restrictive laws of various kinds, but sometimes feelings run high and oystermen take matters into their own hands, even to the extent of armed conflict. As recently as March 1971 such a series of incidents occurred (Huth and Cohen, 1971) when the Maryland Court of Appeals reversed a law of long standing which stated that watermen could work only the waters of their own counties. This immediately brought a fleet of oyster boats from Somerset County, where oysters on the public grounds have suffered mortality from *Minchinia*, to the public grounds in Ann Arundel County, which have been rehabilitated by the state's massive public oyster program. To avoid conflict the governor of Maryland asked the General Assembly to give extraordinary powers to the secretary of natural resources.

It appears that reasonable solutions to the problems of the oyster industry will come about only by the route that so many fishery solutions take—when resistance to change is weakened by disaster and virtual destruction of the industry and the resource. Under these unfortunate circumstances, disease, predation, overfishing, and restrictive laws assume beneficial characteristics, for they or their end effects are the stimuli which prepare people's minds and create receptive constituencies. The oyster industry offers an excellent case-history study to demonstrate how reform in the fishing industry usually comes about only through disaster.

Other Fisheries

Despite the decline in production, oysters are still the leading commercial fishery product in Chesapeake Bay in landed value. The other important invertebrates are blue crabs and clams. The most important finfish, by weight and by value, is the menhaden. Oysters, blue crabs, and menhaden together make up about 80 percent of Chesapeake Bay landings, by weight or by value. Under the Jeffersonian form of government the blue crab and menhaden resources and their fisheries have had interesting histories, which, through their contrasts and similarities, offer some lessons for fishery management. Similar examples can be cited for most segments of coastal fisheries in the United States.

The Blue Crab Fishery

Blue crabs are caught the year round in Chesapeake Bay by a variety of gears (Van Engel, 1962). Most of the catch is taken in four kinds of fishing gear—pots, dredges, trotlines, and scrapes—but crabs are taken also in pound nets, dip nets, or by hand. This year-round fishery, employing a variety of ingenious devices, taking crabs in all conditions from newly-molted soft crabs to females bearing a "sponge" or mass of developing eggs on the abdomen, actively swimming crabs in summer and dormant "hibernating" crabs in winter, is viewed with alarm by many people, including crabbers, crab processors, and recreational crab fishermen who undoubtedly also take large numbers. Yet, despite major fluctuations in abundance from time to time (McHugh, 1969a) the total catch has followed a rising trend for the past eighty years. The period from 1961 to 1970 inclusive has produced the greatest average annual catch in recorded history, although the catch in 1968 was poor.

Scientific studies of the Chesapeake Bay blue crab resource and its fishery have delineated rather completely the complicated and interesting life history of the species (Van Engel, 1958; Cargo and Cronin, 1951) but have brought forth no evidence that the resource ever has been overfished. Indeed, up to now the evidence suggests that the environment, rather than the fishery, has determined the future abundance of blue crabs. It is even possible, although not demonstrated, that eutrophication or enrichment of the estuaries with nutrients from a variety of human sources has stimulated blue crab productivity, as it may have done also for striped bass (Mansueti, 1962; McHugh, 1969a). This is not cause for complacency, for although moderate enrichment can improve biological productivity of living resources useful to man, the process can proceed too far. In the middle Atlantic states, Delaware, New Jersey, and New York, for example, blue crab landings also rose substantially, perhaps from the same cause, but the trend reversed about a decade ago, and there is now virtually no blue crab fishery in any of these states. Nevertheless, the available data on catch and fishing effort for Chesapeake blue crabs support the view that relatively small spawning stocks produce on the average relatively large recruitment and *vice versa*, in a "feedback" mechanism like that demonstrated by Beverton (1962) for North Sea plaice.

The level of research effort on blue crabs has been far from satisfactory, however. When the fishery is doing well there is no support for scientific research, and when crabs are scarce the popular solution is restrictive legislation. In 1968, when crabs were so scarce in Chesapeake Bay that the matter was considered worthy of congressional hearings, Maryland had terminated her blue crab research program and the program in Virginia was minimal (McHugh, 1969a). The evidence to date offers no support for limiting the catch of blue crabs, but the evidence is not conclusive. An adequate and continuing program of research is needed. The results could be of value to all fisheries.

Review of newspaper articles, state and federal legislative hearings, and the like, would lead to quite different conclusions. Public opinion on the condition of the blue crab resource in Chesapeake Bay waxes and wanes in inverse ratio to the abundance of crabs. No one wants to limit his own harvest. He sees calamity in the efforts of others. Thus, there is strong opposition to capture of "sponge" crabs, pressure for limitations on the numbers of pots that can be fished, and arguments for prohibiting the winter dredge fishery in Virginia. Within each state, operators of one type of gear seek to limit the others. Between states, especially when Maryland crabbers consider the Virginia fishery, there is agitation for restriction. The complaints of Marylanders about Virginians taking sponge crabs, fishing too many pots, and generally having no concern about conservation are stimulated by the demonstrable fact that when crab abundance is low, Maryland takes a smaller proportion of the total Chesapeake crab harvest (Cargo and Cronin, 1959). This may be a natural consequence of the migratory habits of the species, but it does not help the situation. The fact remains that Maryland landings fluctuate more widely than Virginia landings, and that this convinces people that the Virginia crab fishery is the root of all problems of the crabbing industry in Maryland.

The usual objective of fishery management is to maintain the resource at a level of abundance which will produce the maximum sustainable yield. With a short-lived animal like the blue crab, which fluctuates widely in abundance from natural causes, the principal goals of research and management should be to understand the environment of the animal, preserve the quality of that environment, and determine the minimum spawning population necessary for effective reproduction.

The life history of the blue crab in Chesapeake Bay and the seasonal patterns of the fishery offer a ready-made method on monitoring the condition of the resource. Most mating takes place in August or September (Van Engel, 1958) in the upper parts of the bay and its tributaries. The male transfers sperm to the spermathecae of the female in spermatophores. Egg laying does not occur until the following May or June. Meanwhile, the females have migrated to the deeper, saltier waters of the lower part of the bay, or into the adjacent ocean. They pass the winter more or less dormant in the deeper channels. Here, the Virginia dredge fishery operates in winter. If crabs are reduced in abundance by the fisheries to such a level that the success of spawning is affected, this might be detected by examining samples from the winter dredge catch to determine the success of mating. Or if females that are not impregnated fail to migrate then the sex ratio in winter in low-salinity waters would merit investigation. Such studies would appear to be important elements in understanding the dynamics of blue crab populations in Chesapeake Bay.

If the conclusions suggested above are valid it would follow that most, if not all, of the laws and regulations with respect to crab fishing in Chesapeake Bay are unnecessary.

The Menhaden Fishery

The fishery for menhaden (*Brevoortia tyrannus*) along the Atlantic coast of the United States is one of our oldest American fisheries. For some reason, probably related to the cost of catching fish, the menhaden fishery did not blossom until after the Pacific sardine fishery developed, created markets for fish meal, oil, and solubles, then collapsed. Thus, the menhaden fishery did not begin to reach its full potential until the sardine fishery was declining in the late 1940s.

Menhaden are caught almost entirely by purse seines, the principal gear used in the United States for densely schooling, pelagic fishes. The life history of Atlantic menhaden and the history of the fishery have many similarities to the life history of the Pacific sardine and the rise and decline of its now defunct fishery (McHugh, 1969b). At the height of the Atlantic menhaden fishery most of the catch was made in the middle Atlantic region (New York to Delaware inclusive), although substantial catches have been made for years in the Chesapeake Bay region and southward. But as the stocks declined in abundance from natural causes, and almost certainly also from overfishing, the proportion of the total catch taken in Chesapeake Bay has been rising, and the proportion taken in the middle Atlantic coastal region has fallen. In the period 1950–62 inclusive over half the total Atlantic coast catch was made off the middle Atlantic coast, and only about one fifth in the Chesapeake region. After 1962 these proportions were reversed and the menhaden fisheries north of Chesapeake Bay now are virtually defunct. The purse seine catch of menhaden in Chesapeake Bay consists entirely of immature fish. The best estimates of maximum sustainable yield of menhaden on the Atlantic coast in the 1950s and 1960s lie between 400,000 and 500,000 tons. In the ten-year period 1953 to 1962 inclusive the average annual catch was more than 658,000 tons.

The menhaden fishery has proceeded with only minimal regulation and with virtually no coordination between states. The only constraints in Virginia have been a six-month closed season at a time when menhaden are scarce or absent, a minimum mesh size for purse seines, and prohibitions on catching menhaden with purse seines in certain tributaries. Maryland prohibits purse-seining of menhaden for reduction, and her catch for bait is only a small fraction (about 1 percent) of the total menhaden catch in Virginia. Because menhaden are highly migratory, this restriction places no limit on fishing effort in the Chesapeake area as a whole. The only limitation in Delaware Bay appears to be a minimum mesh size for nets. Other than license fees, as required by all states, and limitations on fishing for menhaden in some specified inshore waters, New Jersey and New York have no restrictions.

The principal constraint on menhaden fishing has been economic. When the catch drops to the point that operations are no longer profitable fleets are laid up or go elsewhere and plants shut down or at best operate sporadically. Most of the menhaden factories from Delaware north have been closed or have operated only occasionally since 1965, or have used other species of fish. In Chesapeake Bay the catch has not dropped very much, but it has been maintained only at the expense of a substantial increase in fishing effort, although many people in the industry deny that effort has increased.

More and more opposition to menhaden fishing has been raised by recreational fishermen, principally on the grounds that menhaden are the principal food of sportfish species like striped bass, or that menhaden purse seines also catch or scare away sport fishes. These complaints probably are not legitimate bases for outlawing or restricting the menhaden fishery. Sport fishermen usually choose to ignore the fact that most carnivorous fishes, including striped bass (Raney, 1952), feed on a wide variety of fishes and other organisms and probably take menhaden frequently because menhaden often are the dominant species available. This question could be answered more positively if adequate studies had been made of the interactions between menhaden and other filter feeders. But no one knows what species, if any, have captured the energy released by the decline in menhaden abundance. Such information would also be useful to the menhaden industry now, as a clue to alternative resources to maintain the industry. But apart from a few spasmodic and apparently unsuccessful attempts to catch species like sea herring (*Clupea harengus*), river herrings (*Alosa* spp.), and mixed groundfishes, no serious attempts have been made to broaden the resource base.

The menhaden industry might have profited from a study of the history of the Pacific sardine fishery, for the two fisheries have remarkable similarities. But the attitude has been much the same as the attitude of the sardine industry twenty years ago: belief that the decline in abundance is only temporary, belief that the scientific evidence is not conclusive, resistance to regulation by state legislators or fishery agencies, and belated appreciation of the merits of a broad resource base and flexibility in fishing techniques. The individual state legislatures have been unable or unwilling to cope with the problem of rational management of the fishery, and interstate organizations, like the Atlantic States Marine Fisheries Commission, have been no more successful.

Conclusions

Brief historical reviews of the oyster, blue crab, and menhaden fisheries of the Atlantic coast of the United States suggest that despite continued concern about overfishing, pollution, and other human effects upon the environment, no effective management schemes have been established, with one possible exception. That exception is the public oyster management program in Maryland. If one accepts the premises upon which this program is based, it has been successful in the sense that it has about doubled the oyster harvest of the state since the low year of 1963. Maryland oyster landings appear to have stabilized at about fifteen million pounds in the period 1967–70 inclusive, and since 1966 Maryland has been the leading oyster producer in the United States (about 30 percent by weight and 33 percent by value in 1970).

A report to the Wye Institute of Centreville, Maryland (Quittmeyer et al., 1966), examined alternative approaches to the oyster management problems of the state, all of which alternatives included private leasing, and recommended a policy of public management with encouragement of private leasing. The independent research group which made the study included a biologist, a sociologist, and experts in government, business administration, and economics. The present state policy of public management was given low priority, largely on the grounds that the benefits did not justify the cost. It is unfortunate that this research group was restricted to an examination of the fisheries of Maryland, otherwise it almost certainly would have recognized the even greater benefits of removing the social-political barriers to development of a bay-wide oyster industry. Jeffersonian doctrines certainly have not been in the best interests of the Chesapeake Bay oyster industry. Scientific and technological knowledge is available to increase the oyster harvest of the Bay several fold, and to do it in a relatively short space of time, if institutional barriers could be removed and if markets exist for the product.

The Potomac River Fisheries Commission, created by interstate compact to manage the joint fishery resources of the two states, could have pointed the way to enlightened management of the oyster industry and perhaps other fisheries based on endemic resources. Unfortunately, this commission was given neither the authority, nor the freedom from unilateral veto by either of the state legislatures, that it needs to achieve the original concept of its objectives.

The oyster, blue crab, and menhaden fisheries resemble each other in that the living resources on which they are based fluctuate widely in abundance from natural causes. These fluctuations in themselves have aggravated the social-political situation because they create serious economic problems at times for the industry, and because they are not distinguished by most people from the effects of fishing or other human activity. Yet, although against this background of fluctuating abundance the general trend of oyster and menhaden production has been down, blue crab landings have followed an upward trend for eighty years, from less than twenty-five million pounds near the end of the nineteenth century to almost one hundred million recently. The reasons for this upward trend are not known, but the blue crab is among the most popular seafoods of the region and it has been heavily fished for decades. Moreover, unlike the oyster and menhaden, the blue crab has supported recreational fisheries in addition to the commercial catch, and it has been well demonstrated that recreational marine fishing effort on all species has been growing steadily. Under these circumstances it is tempting to conclude that the resource has been increasing in abundance and that nutrient enrichment is the most likely cause. The blue crab spends its early life in those parts of the estuary where enrichment could increase biological productivity, as does the striped bass, which also has produced increasing catches for some forty years. But both these conclusions are pure speculation. There is no objective evidence that abundance has indeed been increasing, and if it has been, we do not know why. Intuitively it can be expected that if eutrophication is the cause we should be concerned about it, for the benefits of uncontrolled and increasing nutrient enrichment could quickly turn into disaster. The story of the oyster industry in Great South Bay, Long Island, is proof enough of that (Redfield, 1952), and the rise and decline of the blue crab fisheries of the states north of the Chesapeake, where the human population is much larger, suggests that disaster may not be far away. Nor should there be cause for complacency in the knowledge that no evidence of overfishing of blue crabs can be found. The catch obviously cannot continue to rise indefinitely, and there is at least a suggestion in the data that as the catch has risen the fluctuations in abundance are becoming greater. This is a common phenomenon in heavily exploited fisheries and it does nothing to improve the economic health of the industry.

The principal effect of Jeffersonian philosophies upon the blue crab fishery has been an abiding faith in the power of legislation, and inadequate and uneven support of scientific and socio-economic research. Each major dip in the catch, which has come at roughly twenty-year intervals, has stimulated restrictive legislation of one form or another. The resource recovered each time, therefore restrictive laws must be good! Science has not found solutions to the periodic fluctuations in abundance, therefore scientific research on blue crabs is not recognized by the industry or the public as important.

Viewed in any light, the history of the menhaden fishery is a prime example of the futility of local control of fisheries on a highly migratory resource. At the peak of the menhaden fishery on the Atlantic coast seven states had annual landings of twenty-five thousand tons or more. At one time or another all fifteen Atlantic coast states have participated in the menhaden fishery. The disastrous history of the Pacific sardine fishery was available as an example of what should not be done. Yet the menhaden fishery, in almost every detail, has followed exactly the same course. It is interesting that the menhaden industry, recognizing the need for good public relations, has adopted a code of ethics to be followed by their fishery vessel crews. The code includes such things as noninterference with sport fishing operations, courtesy when fishing near bathing and recreational areas, and restrictions on fishing at weekends and on national holidays. Some of the positive actions which have come out of this policy are to give menhaden to sport fishermen for bait, and to donate old fishing vessels to be sunk and used as artificial reefs for sport fishing (Hickman, 1969).

As examples of the inability of the Jeffersonian form of government to manage major fishery resources, it would be difficult to find better illustrations than the three fisheries described. As case histories they merit detailed study. The primary lesson they can teach is that strong, well-informed, and centralized control probably is the only effective mechanism for fishery management. When fishery decisions must be made in a regime based on Jeffersonian democracy, special interest groups and the uninformed majority take over and scientific management for maximum yield is impossible.

Various solutions are possible, but their feasibility must be judged in terms of their political acceptability. Federal control is one possibility. An interstate body with plenipotentiary powers is another. Under the status quo public education would be helpful, but the success of a public extension service would depend upon the receptivity of the audience, the adequacy of the background knowledge, and the nature of the fisheries and the resources on which they subsist. The oyster industry would benefit most readily from extension-type activities because grounds could be selected for their capacity to demonstrate the economic advantages of scientific management. With relatively uncontrollable wild stocks like blue crabs and menhaden the value of scientific management is more difficult to demonstrate and arbitrary centralized control is probably the only effective management technique.

April 1971

REFERENCES

Andrews, J. D. 1955. "Notes on Fungus Parasites of Bivalve Mollusks in Chesapeake Bay." *Proceedings of the National Shellfisheries Association, 1954,* 45:157–63.

———, and W. G. Hewatt. 1957. "Oyster Mortality Studies in Virginia, II: The Fungus Disease Caused by *Dermocystidium marinum* in Oysters of Chesapeake Bay." Ecological Monographs, No. 27, pp. 1–26.

Beverton, R. J. H. 1962. "Long-Term Dynamics of Certain North Sea Fish Populations." In E. D. LeCren and M. W. Holdgate (eds.), *The Exploitation of Natural Animal Populations,* pp. 242–64. New York: John Wiley and Sons. Pp. 242–64.

Cargo, G., and L. E. Cronin. 1951. *The Maryland Crab Industry 1950.* Maryland Board of Natural Resources, Department of Research and Education, Chesapeake Biological Laboratory Publication No. 92. 23 pp.

———. 1959. "The Abundance of Crabs in Chesapeake Bay." Maryland Board of Natural Resources, Department of Research and Education, Chesapeake Biological Laboratory Reference No. 59-41. 8 pp. Mimeographed.

Dewey, John. 1940. *The Living Thoughts of Thomas Jefferson.* New York: Longmans, Green and Co. 173 pp.

Engle, James B. 1970. "Oyster and Clam Management." In Norman G. Benson (ed.), *A Century of Fisheries in North America,* pp. 263–76. American Fisheries Society, Special Publication No. 7.

Evans, G. L. 1943. "Story of the Sea-Rac." *The Commonwealth,* 10:10–12.

Galtsoff, Paul S. 1964. *The American Oyster.* U.S. Department of the Interior, Fish and Wildlife Service, Fishery Bulletin 64. iii + 480 pp.

Haskin, H. H., W. J. Canzonier, and J. L. Myhre. 1965. "The History of 'MSX' on Delaware Bay Oyster Grounds, 1957–195." *Annual Report of the American Malacological Union for 1965,* pp. 20–21.

Haskin, H. H., L. A. Stauber, and J. G. Mackin. 1966. "*Minchinia nelsoni* n. sp. (Haplosporida, Haplosporidiidae): Caustive Agent of the Delaware Bay Oyster Epizootic." *Science,* 153 (No. 3742): 1414–16.

Haven, D. 1962. "Seasonal Cycle of Condition Index of Oysters in the York and Rappahannock Rivers." *Proceedings of the National Shellfisheries Association, 1960,* 51:42–66.

Hickman, Milton T. 1969. *Seventieth and Seventy-first Annual Reports of the Marine Resources Commission for the Fiscal Years Ending June 30, 1968, and June 30, 1969.* Newport News, Va. 50 pp.

Hopkins, Sewell H., and R. Winston Menzel. 1952. "How to Decide Best Time to Harvest Oyster Crops." *Atlantic Fisherman,* 33, No. 9 (Oct. 1952): 15, 36–37.

Huth, Tom, and Richard M. Cohen. 1971. "Law Urged to Avert Oyster War in Md." *The Washington Post,* Wednesday, March 17, 1971.

Korringa, P. 1952. "Recent Advances in Oyster Biology." *Quarterly Review of Biology,* 27:266–308, 339–65.

Logie, R. Reed. 1956. "Oyster Mortalities, Old and New, in the Maritimes." Fisheries Research Board of Canada, Progress Reports of the Atlantic Coast Stations, No. 65, pp. 3–11.

Loosanoff, V. L., and H. C. Davis. 1963. "Rearing of Bivalve Mollusks." *Advances in Marine Biology,* 1:1–136.

Mackin, J. G., H. M. Owen, and A. Collier. 1950. "Preliminary Note on the Occurrence of a New Protistan Parasite, *Dermocystidium marinum,* in the Barataria Bay Area of Lousiana." *Science,* 111:328–29.

Manning, J. H. 1969. "Bay Fisheries Resources." *Proceedings of the Governor's Conference on Chesapeake Bay, Sept. 12–13, 1968,* 2:91–101.

Mansueti, R. J. 1962. "Effects of Civilization on Striped Bass and Other Estuarine Biota in Chesapeake Bay and Tributaries." *Proceedings of the Gulf and Caribbean Fisheries Institute, 14th Annual Session, Nov. 1961,* pp. 110–36.

McHugh, J. L. 1969a. "Fisheries of Chesapeake Bay." *Proceedings of the Governor's Conference on Chesapeake Bay, Sept. 12–13, 1968,* 2:135–60.

———. 1969b. "Comparison of Pacific Sardine and Atlantic Menhaden Fisheries." *FiskeriDirektoratets Skrifter, Serie HavUndersøkelser,* 15:356–67.

———, and J. D. Andrews. 1955. "Computation of Oyster Yields in Virginia." *Proceedings of the National Shellfisheries Association, 1954,* 45:217–39.

Nixon, R. M. 1971. *State of the Union Message.* Washington, D.C.

Quittmeyer, C. L., J. D. Andrews, G. C. Jones, V. A. Liguori, D. B. Pettengill, and A. L. Sancetta. 1966. *Chesapeake Bay Fisheries of Maryland.* Centreville, Md.: Wye Institute. vii + 68 pp.

Raney, Edward C. 1952. "The Life History of the Striped Bass, *Roccus saxatilis* (Walbaum)." *Bulletin of the Bingham Oceanographic Collection,* 14 (No. 1):5–97.

Redfield, Alfred C. 1952. *Report to the Towns of Brookhaven and Islip, N.Y., on the Hydrography of Great South Bay and Moriches Bay.* Woods Hole Oceanographic Institution, Reference No. 52-26. 80 pp.

Schlesinger, Arthur, Jr. 1971. "Is It Jeffersonian?" *New York Times,* Saturday 30 Jan. 1971.

Sindermann, Carl J., and Aaron Rosenfield. 1967. "Principal Diseases of Commercially Important Marine Bivalve Mollusca and Crustacea." U.S. Department of the Interior, Fish and Wildlife Service, Fishery Bulletin 66 (No. 2): 335–85.

Van Engel, W. A. 1958. "The Blue Crab and Its Fishery in Chesapeake Bay, Part 1: Reproduction, Early Development, Growth, and Migration." *Commercial Fisheries Review,* 20 (No. 6):6–17.

———. 1962. "The Blue Crab and Its Fishery in Chesapeake Bay, Part 2: Types of Gear for Hard Crab Fishing." *Commercial Fisheries Review,* 24 (No. 9): 1–10.

Wells, William Firth. 1922. "Problems in Oyster Culture." in *Early Oyster Culture Investigations by the New York Conservation Commission* (1920–

1926), pp. 17–31. Reprinted 1969 by State of New York Conservation Department, Division of Marine and Coastal Resources, Ronkonkoma, N.Y. Pp. 17–31.

Yonge, C. M. 1960. *Oysters*. London: Collins. xiv + 209 pp.

9

The Gap between Theory and Policy in Fishery Development

WILLIAM F. ROYCE

The difficulties of applying both biological and economic models to real fishery problems, particularly in a rapidly changing situation, are mentioned by a number of contributors to this volume, notably James Crutchfield. The biological models usually depend on describing the population dynamics of a fish stock in an equilibrium situation, and therefore the models are really describing "population statics." The economic models depend basically on good biological models, but they incorporate some additional difficulties of application, especially with respect to methods of measuring benefits and to disparities between the units of fish stocks and the units of fishing operations.

My concern here today is with two major, interrelated problems. The first of these is the kind of objectives. These can be biological, for example, a maximization of yield; or economic, for example, a maximization of benefits; or political (the area that has tended to be overlooked by academicians), for example, assurance to a large group of people that certain things are being done. The second major part of my thesis involves the diversity of the fisheries. We have tended at the University of Washington to think predominantly in terms of the food fisheries, but the United States public is aware of many other kinds of fishery problems that determine to a large extent governmental actions.

The gap between theory and reality will be approached through a discussion of the problems of priority in development of various kinds of fisheries. The term "development" is used rather than "management" because it implies the continuing improvement and enlargement of our fisheries rather than the management of a status quo. In a general sense any of our fisheries can be considered to be underdeveloped because stocks are either underexploited or overexploited, or because the en-

William F. Royce is associate director for research of the National Marine Fisheries Service. At the time this paper was presented he was associate dean of the College of Fisheries, University of Washington. He has been a director of fishery research activities and an advisor on fishery problems in many parts of the world.

vironment of the stocks can be improved, or because there is reason to change the distribution of benefits.

When we use the word *fisheries* in the general sense, we need to consider the meaning of the word to the general public, yet there has been no careful examination of public attitudes concerning the word. The circumstantial evidence suggests that we ought to consider separately four categories of fisheries: the recreational fisheries that are sustained in part by public aquaculture; private aquaculture; the inland and inshore commercial fisheries; and the distant water commercial fisheries. These groups of fisheries vastly differ in method of exploitation, in the mechanics of governmental actions, in public attitudes, and in problems of priority.

Public Interest

In the absence of an empirical study on the public attitudes toward the fisheries, we must infer the attitudes from a variety of evidence. The controlling circumstances are probably the common-property status of the fish, a broad public recognition that many fish stocks have suffered depletion or even extinction, and the current interest of people in fishing.

Uppermost in the minds of many is probably the fact that each person has a right to the fish. He feels that he owns part of the resource ("a piece of the rock," in the language of the current Prudential Insurance advertisement). He enjoys a privilege that is now nearly universal but was not always so because in earlier times some of the choice fisheries were reserved to royalty or the wealthy. The common-property status has been a hard-fought gain and is not one to be easily relinquished.

Second, the practice of fishing brought evidence long ago of the ease with which the resources can be depleted. Especially significant is the extinction or decimation of the salmon in some European streams in the eighteenth century. Stocks of numerous freshwater fishes were badly depleted in the eighteenth and nineteenth centuries also. These highly visible results of greed the average person does not want repeated with his resources. He probably regards the concept of maximum sustainable yield as just an excuse to deplete other resources.

Third, the role of the fisheries is rapidly changing in many of the more advanced countries. A century or two ago most of the internal and inshore waters were fished for food by people who were otherwise engaged in agriculture. The production of fish was a supplement to the production from the land, and a large part of it was consumed by the fisherman's family or sold fresh in the neighboring community. (Such practices continue at present in many of the lesser developed countries of the world.) For example, in the United States some reports of fishery departments and fish commissions in the nineteenth century show tabulations of the gear and catches from commercial fisheries that existed on a large proportion of the inland waters. Included in these catches were large numbers of species that are now commonly designated as game species, such as trout, bass, pike, catfish, and others.

The role of these internal and inshore fisheries has shifted from a source of food to a major component of our outdoor recreation in North America. About 25 percent of the population in North America now participates in recreational fishing and is willing to spend a large amount of money to do it. In the United States the ratio of recreational fishermen to commercial fishermen is about 350 to 1. The total catch of fish suitable for direct human consumption by recreational fishermen exceeds that by commercial fishermen.

The individual attitude toward recreational fishing is highly varied. Some people find satisfaction in just fishing on the bank of a quiet pond. Others may be anxious to catch a bag full of fish to eat. Still others may travel half around the world and spend tens of thousands of dollars in an arduous search for a special trophy. Regardless of the kind of recreation involved, most people cherish the opportunity to go fishing. Even if they do not go, they want to know that their resources are in good condition and available if they should want to go fishing. When they do, they want to know that they have as good a chance to catch fish as the next person. Bitter are those who try to fish fairly and legally when they see someone catching fish illegally. Anglers frequently feel antagonistic toward commercial fishermen even when the commercial fishermen are not catching any game fish.

The practice of recreational fishing is, of course, utter nonsense from the concept of efficiency. Instead there is a concept of sportsmanship in which a person uses his individual skill in a way that may prove he is a bit more skillful than his fellows. He may use an absurd lure and a tiny leader and operate them with the most efficient possible rod and reel. He will not tolerate any team fishing that might involve the use of a net or anything destructive, such as explosives.

When we consider that this kind of fishing is practiced by about 25 percent of our population and commercial fishing for a livelihood by less than one tenth of 1 percent, we realize that the general public attitude toward fishing derives mainly from its experiences and concern for the recreational fisheries. The recreational fisherman is not concerned with efficiency. And why should he be worried about the efficiency of a commercial fisherman, especially when that commercial fisherman might reduce his opportunity to go fishing for fun?

Development Problems

International Distant Water Food Fisheries

These fisheries are defined as those conducted outside the territorial seas or contiguous fishing zones of coastal nations. Most of the vessels are large, expensive, technically sophisticated craft that can operate a long distance from port for long periods of time. Usually they operate as part of a fleet and as such can fish any ocean in the world. They are owned by private, profit-oriented operators, who are commonly subsidized by governments or owned by countries with centrally planned economies. There is no aquaculture in these fisheries now, nor is there likely to be in the future.

Almost all of the resources sought by these fisheries are outside any significant institutional control. They are generally available without restriction to fishermen who can find and catch them. Such resources have been especially attractive in recent years to the long-range fleets that have had the capability to move quickly from one area to another. In a number of areas these fleets have demonstrated their ability to reduce the abundance of the resources grossly within three years, so rapidly in fact that the coastal countries that might have been concerned were scarcely aware of the scope of operations until too late.

The operations can be fairly called "search and destroy operations." The operators and their governments adhere vehemently to the principle of the freedom of the seas and few have accepted any responsibility for maintenance of the resources.

The urgent problem in the development of these fisheries is the establishment of effective control so that overexploitation of the stocks is restrained and yet the discovery of new stocks or the creation of innovative fishing methods is not inhibited. A modest amount of control has been accomplished by some multilateral and bilateral agreements or by extension of the authority of the coastal states. The latter method is, in practice, much more effective.

A second problem of high priority is the need for much better information on the locations of these resources, the levels of sustainable yields, and the consequences of alternate strategies of management.

National Food Fisheries

These fisheries are pursued in inland waters, territorial seas, and contiguous fishing zones by a mixture of large, expensive vessels and relatively small, primitive equipment. The operations in countries without centrally planned economies are all profit-oriented but subsidized by the governments in practically all instances. The large vessels may take part in distant water fishing as well.

From the mixture of large, sophisticated and small, simple equipment comes, first, a different set of political problems than that of the international distant fisheries. The small, simple equipment is usually operated by large numbers of traditional fishermen, who cannot easily change their way of life or overcome their lack of skills, their lack of money, or other difficulties. Any attempt to reduce costs in these fisheries by favoring the better equipment or by reducing the numbers of fishermen will produce urgent secondary problems of unemployment and dislocation of people.

Second, the welfare of these fish stocks, except for those stocks that wander in and out of national waters, is plainly a responsibility of the government. The dual concern for the welfare of the resource as well as that of the fishermen creates a strong political pressure toward inefficiency. There is a tendency to force large, efficient vessels away from the operating area of the primitive fishermen into the international fisheries. In those national fisheries where yields have reached the maximum sustainable levels, the fishermen face great difficulty in keeping costs down and earnings and return on capital adequate. Nevertheless, the improvement of such inefficient fisheries requires raising the efficiency and reducing costs if private investment is to have a part.

On the other hand, in practically all countries there are natural stocks that are underexploited. The governments can help the private sector by planning and operating port facilities, by designing more efficient vessels, by building pilot plants that will reduce development costs, and by helping with marketing. They may also train fishermen or other workers and establish sanitary or quality standards. These activities are really what we normally think of as development activities and they require an intimate knowledge of the complex of problems that arise from the underlying needs of protecting the resources and providing for an adequate return to labor and capital.

Recreational Fishery Development

People have fun with fish through sport fishing, fish watching, and keeping fish as pets. The sport fishing and fish watching opportunities in North America are provided predominantly by governments, but the pet fish industry is almost entirely private.

Sport fishing is only a part of a broader recreational experience involving travel, outdoor living, hiking, and boating although it may be the primary excuse for a complex activity.

The development of recreational fishing involves the control of pollution, provision of access to fishing areas, management of new waters, notably reservoirs, better management of existing waters, and the development of varied transportation systems from highways to hiking trails. The actual management of the fish populations is an important but relatively minor part of this activity.

A large part of the government support of recreational fishing is devoted to facilities for public fish culture. Most of this kind of fish culture is for the purpose of providing trout of a catchable size just before and during the fishing season. A more limited part of it is involved with controlling the environment and moving fish to new waters either by transferring large individuals or by hatching the eggs and shipping the fry.

A substantial and private part of the sport fishing complex is the provision of bait. The golden shiner, which is the principal bait fish, is probably grown in larger numbers in North America than any other species of fish.

Not to be overlooked is the pet fish industry, which is private and so secretive that good statistics are not available. It does appear, however, that more than twenty million households keep aquaria and that the trade handles about five hundred million fish annually in addition to a large volume of equipment, food, books, and so on. The total sales volume of pet fish and related supplies may well be close to that of food fish in the United States.

Private Aquaculture

Great prospects have been assumed by many people for this fishery, but these are almost surely a long time in the future. The activity is a part of our food industry that is entirely profit-oriented.

Aquaculture, like agriculture, is a circumvention of the natural controls on populations. We control the numbers of animals, their breeding and genetic composition, their nutrition at all stages of life, and their predators, competitors, and diseases. Our scientific advances for exercising these controls in aquaculture have lagged far behind those made in agriculture.

In addition to the great need for solid information on nutrition, disease control, and the breeding of animals, the private aquaculturists who are utilizing public waters are subject to special problems. They usually operate under the control of the fish and game departments and they are frequently accused of polluting the waters or interfering with other public uses. In fact, their operations are subject to almost continual harassment by the fish and game departments or other public agencies in many states.

The private fish farmers who operate their own ponds face a quite different set of legal-technical problems. Highly important is the clearance of suitable disease control agents, many of which have already been cleared for use on farm animals but which must be cleared again for use on fish. Such clearance commonly involves a quarter-million-dollar study and is beyond the capability of small operators. A second major problem is the control of disease transfer. The operator is, of course, anxious to sell his fish and at the same time to avoid bringing any diseases in. Another related problem arises from the cultivation of exotic species that may not occur naturally in neighboring waters. Most states now have laws prohibiting the introduction of exotic species because of fear of their accidental release.

There are encouraging trends in the United States with respect to a shift of responsibility for some aquaculture over to the Department of Agriculture, which is now taking a strong interest in fish farming and providing much more assistance than the fish farmers could obtain from the fishery agencies.

Comment on a Theoretical Problem

Earlier in this paper I mentioned that our procedure for examining yield functions under stable conditions had been referred to as "population statics" rather than "population dynamics." This criticism is not too serious because a number of people are making great strides toward a more pragmatic approach through the use of such models. I do want to make the point, however, that we need a more general consideration of yield functions.

The familiar case is the positive relationship between effort and yield that has a maximum, or an asymptote. This function is commonly applied to the food fish stocks and may be extended to a few of the recreational stocks in which the quantity of fish caught is of primary importance to the fisherman.

Our theories ought to include an opposite extreme in which any production will have a negative effect although the fishing effort is useful. Let us consider the management of the Loch Ness Monster. Whether this is a real or imaginary animal is beside the point; it is a significant resource to the people of the area who benefit from the legend. It attracts many visitors, including the so-called scientists who fool around with sophisticated gadgets. If this resource should ever be found or explained it almost surely would depreciate in commercial value; and if I were managing it, I would take great pains to keep the legend going indefinitely.

A fish stock with a similar negative production function might be the animals on a coral reef that is designated as an underwater park. These, as well as the Loch Ness Monster, are of special interest in their existing condition which should be maintained for nonconsumptive uses.

We might reject such examples as ridiculous extremes that are not really a part of fish population analysis except for the fact that we seem to have a complete intergradation between the negative and positive yield functions. We have recreational fisheries in which the catch is in fact almost incidental to enjoyment of the wilderness. For example, I met a man recently who expounded on his enjoyment of Atlantic salmon fishing. He felt that it was the finest sport that he had ever engaged in. He had made several trips to his salmon club in Nova Scotia and thoroughly enjoyed them although he had yet to catch a salmon. To use an obvious metaphor, his enjoyment came from "nibbling at the lure."

March 1971

10

Fisheries and the National Interest

WILLIAM M. TERRY

This paper deals with the future of the fisheries of the United States from the point of view of broad-based strategy rather than of operational detail. In developing this topic I shall speak as a government official, not as a representative of industry or the academic community, not as a scientist, or an economist, or a technologist, or a law enforcement expert, but rather as an administrator responsible for spending the general revenues of the United States with, therefore, the general interest in mind as well as the particular.

The first thing I have to do in approaching the "future and broad-based strategy" is ask a question which to people in the fraternity may appear to be heretical. The question is, "Why should government treat the fishing industry any differently than it treats lots of other industries, the hula hoop industry for example? That is, why should the general revenues be spent on the fishing industry?" A lot of answers have been given; let me run through four or five of them. One answer is always, "Well, one must always take care of the fishing industry because it produces food." But in a country in which the per capita consumption of fish as direct food is eleven pounds and the total consumption of other things as direct food is in the neighborhood of fifteen hundred pounds, can one really consider fish very important?

Another answer is, "Because of the balance of payments problem. That is to say, the American people like to eat fish and they will acquire it wherever they can and at the moment they are acquiring three quarters of it, more or less, from outside the country at the cost of a serious imbalance in trade. We should foster a U.S. fishing industry in order to

William M. Terry is presently director of international affairs, National Oceanic and Atmospheric Administration, United States Department of Commerce. He has participated, on the behalf of the United States, in most of its important fisheries negotiations since the 1950s, serving as either principal advisor to the delegation or as spokesman for the delegation; he also participates as United States representative to major fisheries commissions, such as the International North Pacific Fisheries Commission, the International Commission for the Northwest Atlantic Fisheries, and the Inter-American Tropical Tuna Commission. He has served as acting deputy director of the Bureau of Commercial Fisheries and director of the Office of Foreign Activities in the Fish and Wildlife Service.

remove that imbalance." The problem here is that the fish, by and large, which come into the country come from three countries, Mexico, Canada, and Japan. The United States, for its own well being, is committed to supporting the economies of those three countries. This means that we must buy something from them. If we don't buy fish we buy automobiles or what have you—wheat from Canada, tequila from Mexico, except we probably cannot consume that much, and rice from Japan, although we have lots of rice right here. This is not to say that it might not be better to buy wheat from Canada. It is to say that to insist upon the development of the fishing industry to solve the balance of payments problem is an oversimplification.

Another answer that is made frequently is about the need for a fishing fleet during the times of a national emergency. "The U.S. in wartime," for example, "needs a fishing fleet for defense." My difficulty with this answer is that I have never heard a naval officer make the argument; it has always been a fish expert who has done so.

Still another argument is that fishing is an economic activity which provides opportunities for employment, and the government should promote the industry in order to increase the opportunities. But one wonders how this really applies in today's United States. One wonders if the normal criteria of the market should not control.

Finally, we hear a great deal about leadership, we hear a great deal of weeping and wailing and gnashing of teeth about the fact that the United States has fallen from second place to third place to sixth place or fifth place among fishing nations. I have never been able to understand the relevance of this argument, particularly when we are being compared with such countries as the USSR, in which there is direct control of the economy, and in which there is an extraordinary motivation for the allocation of resources to exploitation of the sea.

This begins to sound as if there is no reason why government should treat the fishing industry any differently than it treats the hula hoop industry. But that is not the case, I think. There are several possible points of contact or reasons why there should be government intervention in the market. The first of these relates to those instances in which one part of the private sector, in this case the fishing industry, is disadvantaged by activities of the government undertaken for the benefit of other parts of the private sector, for example, when government activities which seek to advantage the farmer result in giving him an unduly advantageous competitive position over the fisherman.

A second instance in which government intervention seems warranted is when the ordinary criteria of the market result in a misallocation of the resources, that is, in an economic imbalance, the effects of which are felt in a substantial way outside of the fishing industry itself. A third obvious cause for government intervention is for purposes of control for resource protection.

If we go back and look again at the first point, it clearly raises the question of fairness. When the government takes an action which helps one part of the population at the expense of another there is a question of equity involved. The difficulty with this argument in favor of intervention is that it inclines to a political resolution, the idea of the squeaking wheel, and it is, in my judgment, a poor basis for a national fishery policy. The second and third points seem to me to offer an adequate basis for policy. Let me elaborate them beginning with the third point, protection of the resource.

From the national point of view there is a long-term need to preserve options available in coastal fishery resources. These resources may not be important now, and one may, if one chooses, argue that the industry which depends upon them is of no overriding material consequence at the moment, but it would seem to be foolish for government not to preserve options, at least if the cost of doing so is modest. Thus, resource protection and activities necessary to achieve it seem to be something about which a national fishery policy can be built. The government role here is justified because in the absence of property rights in the resource there is no incentive for the individual to take action to protect resources.

The second point I made had to do with market failures, which occur when the ordinary criteria of the market result in misallocation of resources, waste of the nations' resources of men and capital, which produces substantial adverse consequences both in the fishery and outside of it. Here it seems mandatory for the government to intervene in the national interest. In many, if not most cases, the cause of the failure is institutional and thus beyond the range of industry self-help.

In both cases I have tried to identify a national need. It is important to keep this in mind because most of what I have to say from here on relates to this need.

There is one further comment to be made at this point. It is my own opinion that the political realities in the United States are such that we are safe in saying that there will be a federal fishery program. That is to say, there will be expenditures of funds from the general revenues to assist the fishing industry. By assisting the industry I mean going beyond the mere protection of the resource or maintenance of the resource at optimum levels of abundance. These expenditures can attack root problems or they can attack symptoms. In the latter case the result is a sort of perpetual subsidy and constant burden to the taxpayer. In the former case it is possible to rationalize industry, to make it competitive, hopefully to put it on its own two feet and get it off the back of the general taxpayer. It seems fairly obvious that we ought to attack the problems.

What then are these problems? In answering the question I will speak primarily about the producer, that is to say, the fisherman, the boat owner, rather than the processor, the broker, or the fish salesman. The problem is reflected in the fact that during the past twenty years there has been essentially no increase in United States production of fishery products although during the same period of time the demand increased enormously. The demand today is three times domestic production. What is the cause? Basically the causes can be said to be related to high costs. These result: (1) from technological inadequacy or failure on shipboard, in the processing plants, and in the distribution chain; (2) from reduced or uncertain sources of supply, and here I speak not merely of abundance of animals in the ocean but also uncertain access to the abundance; (3) from efficiency-inhibiting laws and regulations, both federal and state; and (4) from government action on behalf of other segments of society, for example, the statute prohibiting the use of foreign-built vessels in the United States fisheries.

The first two of these, however, are symptoms in fact, not causes. We can in the National Marine Fisheries Service do technological research on new products and develop new uses for products. We can engage in exploratory fishing and also in resource enhancement to remove uncertainties about supply. We can engage in marketing activities and financial assistance programs. In reality none of these activities gets at the root problem. Further, we must recognize the fact that much of industry, that is, industry other than the fishing industry, is capable of doing its own technological research and avoiding obsolescence.

"Why can't the fishing industry do this?" one must ask. The answer, of course, is that some parts of the fishing industry do, but that by and large industry does not have the incentive to make the capital investment required to overcome the technological problems. Why? Because of basic institutional constraints.

In regard to the second cause listed above, within limits government can maintain resources at high levels of abundance, but there are clear limits imposed again by institutional constraints.

These constraints of which I speak are international law which results in the absence of property rights on the high seas as between nations; the national political system of the United States which results in the division of jurisdiction over fishery between state and federal governments; and our national tradition which results in the common property character of the fishery resource or the absence of property rights as between individuals within the United States. The last two apply not only to fishery resources but also to the environment in which the fish dwell.

These institutional constraints are in my judgment the root causes, the root problems, the root evils which confront the fishing industry. These create the market failures which produce economic soft spots and the need for assistance. These produce the valid basis for government intervention in the market place in connection with fisheries. These create the obstacles, particularly in the international arena, to United States government efforts to protect the resources in the long-term interest of the nation. These must be solved or the result is permanent subsidy or, in the absence of that, a permanently sick fishing industry.

If we look around the world we find that there are two traditions. One is our tradition of the common property resource and with this tradition we find two sets of circumstances, either a more or less permanently subsidized fishing industry, or a sick fishing industry such as we have in many parts of the United States. The other tradition is that of creating property rights in effect, practiced somewhat in Japan, and practiced in the USSR and other more centrally controlled nations than ours. And here it is not uncommon to see a healthy fishing industry.

The impact of the former tradition is pretty clear. Look at it internationally—common property resources in the high seas. This freedom of the seas means precisely too many fishing vessels from too many countries working on the resource at the same time with the threat of depletion, but usually before that, economic disadvantages from overcapitalization. Our traditional way out of this kind of problem is through international organizations, which do research, make regulations, and apply them, for example, the International Commission for the Northwest Atlantic Fisheries (ICNAF), the Inter-American Tropical Tuna Commission (IATTC), the International Pacific Halibut Commission, and others which are well known. The difficulty is that these organizations were designed for the days of lesser fishing effort, fewer nations fishing, when size limits, mesh regulations, and measures of that type were sufficient. Now with the need for control on fishing mortality in the form of catch quotas or effort quotas, these organizations do not work well in the absence of agreement on allocation of the yield among the competing nations. The two fairly successful international agreements are those related to the development of the International Pacific Salmon Fisheries Commission and the International North Pacific Fisheries Commission, and these are successful largely to the extent that the commissions themselves solve the allocation problem. The former provides for a 50-50 split of the fish, and the latter provides for a 95-5 split of North American salmon, or about that, between North American nations and Japan. The others, in which there is no provision for resource allocation or no provision for obliteration of the common property concept, are struggling. ICNAF is struggling. IATTC is struggling, as is witnessed by the fact that this year's annual meeting took approximately one month, contrasted with a normal annual meeting of two or three days. Both these commissions are faced with extraordinary difficulties. ICNAF has fourteen or fifteen members with greatly differing degrees of sophistication in research and fishery development, with sharply differing political systems, and with great variations in economic development, from relatively underdeveloped Portugal, for example, to highly sophisticated West Germany. IATTC faces the same kinds of problems, but overriding these and fundamental to both commissions is the fact that neither of them attempts to deal with the resource allocation problem. Let me note, however, if we return to the North Pacific Fisheries Commission, that the problem of domestic allocation of the resource and the environment remains.

We also attempt to get at these problems through bilateral agreements, but these agreements, *ad hoc* in nature, are less than satisfactory, mostly because they are *ad hoc* and provide only interim and very parochial solutions.

Let me elaborate somewhat on this question of the allocation of resources in the international arena.

When we concern ourselves with the problems of allocation on the international scene there are constraints which tend to affect what we do and how we do it.

The first constraint is presented by present day international law, which limits national control or jurisdiction over fisheries to twelve miles from the coast as a general rule. There are some exceptions. Some nations claim more, such as Ecuador, Peru, and Chile, but as a rule it is twelve miles and it will not change quickly. International law of course permits *ad hoc* arrangements which may produce a different result, but basically high seas fishery resources are common property resources among nations as well as among fishermen.

The second constraint is in a sense a product of the first. Since the resource is open to all on an equal basis we must deal with all on an equal basis. It is seldom that we deal with another country with exactly the same political philosophy or economic establishment as ours. This means there are few common meeting grounds on which we can come to agreement other than the principle of maximum sustainable yield.

Under these constraints there are two general modes in which we can act. The first is to achieve, via whatever mechanism is available, a designated share of the yield of the resource in question. The total yield level will have the concept of maximum sustainable yield as its basis and that concept may also dictate certain limitations on how the yield is taken. Our share of the yield will be determined in the end by social and economic considerations important to the parties to the negotiation, but there may well be restrictions on how we take it. Clearly it cannot be taken in such a way as to mitigate the principle of maximum sustainable yield. Let me illustrate. In order to maintain or achieve the maximum sustainable yield, it may be necessary to limit fishing mortality in any one year by means of direct catch limitation, and it may be necessary also to exclude a particular part of the potential harvestable population from the catch. A mesh regulation is one mechanism that is used to accomplish the latter objective, since it is designed to insure that fish below a certain size escape capture. In these circumstances we do not have the option of taking our share in any way we choose. We do, however, have the option of allocating it among competing users in the United States in any way we choose.

The second mode in which we can operate is one in which by some mechanism we obtain total control of the resource. The best example that I think of at the moment is that which we obtained when the United States extended its jurisdiction over fisheries to twelve miles. Any resource totally contained within this boundary of course need not be the object of an international fishery.

Certain implications of these two modes are important to note. Generally speaking the quantity of data, the precision of estimates, the extent of understanding required before action is taken to regulate fishing or to allocate the resource are far greater when we operate in the first mode than when we operate in the second. The standard of proof required to be met in the international community is generally speaking higher, for obvious reasons, than that required when a single government has total control. The result is the expenditure of a great deal of effort in connection with, for example, Alaska salmon, North Atlantic groundfish, river herring, scup, fluke, and so on, in the mid-Atlantic bight, ocean perch off the coast of Oregon, than is on the average needed to manage for maximum sustainable yield. This is because in each case we are operating in the first mode and are required to establish certain facts to the satisfaction of the governments of the USSR or Poland, or Japan, or Canada—facts which those governments do not really want to accept.

It is also important to note that for any resources that extend beyond twelve miles, the domestic allocation problem can hardly be approached, let alone be solved, until the international allocation problem is solved.

Our long range objective is to change international law so as to deal more rationally with the allocation problem—that is to say, quite candidly, to change international law in such a way as to create property rights in the resources of the high seas. There are, of course, many difficulties involved, not the least of them residing within the country itself. Let me illustrate the problem. The most obvious one is the fact that two very important United States fisheries, tuna and shrimp, find certain advantages in not having property rights developed on the high seas—at least temporary advantages. At the moment the United States tuna industry has the lion's share of the yellowfin and skipjack catches in the eastern tropical Pacific, likes it that way, and does not want any allocation of resources among competing nations, because the allocation, of course, will come from what is now the United States catch. The United States shrimp industry operating out of the Gulf and South Atlantic ports operates substantially off the coast of Mexico, other Central American countries, and off the northeastern coast of South America. Here resource allocation in the most likely pattern, that is, by extension of jurisdiction, works very much against the interest of the shrimp industry.

A further aspect of the difficulty associated with this problem is the fact that the most popular solution to it, the extension of jurisdiction to two hundred miles or something of that nature, gets us into enormous complications with other interests in the United States, interests which consider the drawing of lines here and there on the ocean surface to be anathema. So, there are very considerable difficulties involved.

We in Washington, D.C., have been working now for some time in preparation for a third United Nations Conference on Law of the Sea in 1973, and have been attempting to develop the concept of coastal state control not expressed in terms of geographic jurisdiction, but rather related specifically to resources. We have in mind, control for conservation purposes which would be nondiscriminatory, resulting in the application of conservation measures to all comers on an equal basis; and second, and equally important, control for economic purposes which would result in giving the coastal state a preference related more or less to the coastal state's needs or special circumstances. Even if the conference is really successful it would be several years later before the rule developed to the point at which it would be generally accepted. This seems to be the best hope for a solution to the international allocation problem.

Domestically we have the same problem—the tradition of free fishing, every man with a God-given right to go fishing, and, therefore, no property rights. The result is that effort is sucked into the resources and we have endless competition with commercial fishermen competing with other commercial fishermen, commercial fishermen competing with sport fishermen. When it becomes necessary to regulate for the purposes of optimum abundance, we end up applying regulations which are inefficiency producers, high cost producers. The examples that I can cite are the sockeye salmon fishery, the tuna quota, and some of the regulations applied to Alaska's salmon, the limit-seiner for example. These things all produce an Alice in Wonderland world in which government, because of the tradition of free fishing, has no control over the number of units of effort which are applied to a fishery, and must, if it wishes to limit fish mortality, seek to make each unit of effort less efficient than would be optimal. I am not talking about state regulatory systems or efforts alone. The problem is equally evident in federal regulatory systems.

It should be very clear here that we cannot have rational management systems until this problem is solved. We have something approaching economic nonsense in our present regulatory systems, a perpetual waste of the nation's resources of man power and capital. If we can solve this, we have a chance at least of putting industry on a paying basis, a basis on which not only will the return on investment be enormously increased as a result of the decrease in redundant fishing effort, but on which there will be the option for the government to rake off, if you will, a little bit of that increased return on investment to pay the cost of government expenditures.

But let us look at the third institutional complication that I mentioned earlier, the domestic political system. Here we have state jurisdiction reaching out on the average to three miles, federal jurisdiction beyond that for a certain distance, and then no jurisdiction or international jurisdiction if you want to call it that. I have suggested a rationale for federal intervention in the market: (1) to protect the nation's options for the future in the national interest, (2) to prevent market failures which have substantial economic consequences, also in the national interest, and (3) to rationalize industry, to offer the possibility of a "pay its own way" industry, also in the national interest. But the reality of the political system is that in large part the state governments have the power to negate entirely the implementation of a national policy, the achievement of national objectives. In reality it seems impossible to have a viable national fishery policy, if thirty or thirty-five coastal states can operate without regard for national objectives. As I have said, the states control for conservation purposes up to three miles and in practice beyond that in many cases. Alaska, for example, theoretically has control over Alaskan citizens alone beyond three miles. As a practical matter, through landing laws and the need for logistic support in the state, most fisheries in Alaskan waters can be controlled by state authorities. In addition, states have control for economic purposes almost totally, leaving aside the interstate commerce clause of the constitution.

The manner in which the yield from the resource or a share of it is allocated determines the economic and social value to the United States of the resource. If the basis for and manner of allocation are rational and fit some national-interest economic and social framework, a national objective is achieved and the expenditure of the national revenues is justified. On the other hand, it hardly seems rational to spend the general revenues on a federal program aimed at creating a certain social or economic impact in a part of the private sector if the program can be negated by the parochial action of a state or states in the allocation process.

It should be noted that both resource protection and the achievement of certain economic and social objectives depend on solution of the allocation problems. We have found, nationally and internationally, that effective allocation of a resource is often a prerequisite to effective measures to insure optimum abundance of the resource.

There is one further point that I have said very little about so far, but it is a point to which everything said earlier applies. It is also an allocation problem, largely a domestic one, but one which will doubtless become an international problem of some consequence. This is allocation among competing users of the environment in which fish live. Many species which support commercial fisheries and an even greater number of species which support recreational fisheries spend a critical part of their life cycles in the coastal zone. The manner in which this "near-shore" zone is used, and its allocation among competing users, the quality of this environment then, are critical. Competing uses for the marine environment result in: (1) lower water quality if the environment is used as a sewer; (2) changes in the physical character of the environment if extensive dredging and filling operations take place; (3) virtual pre-emption if the area becomes populated with oil wells, valves, and pipelines; and (4) pre-emption if exploitative activities are banned in the interest of aesthetic enjoyment of the aquatic environment. The connection between this problem and the resource allocation problem will be immediately apparent; and it will be clear that the solution of the latter without solution of the former is a useless exercise. Not only must there be coordinated and integrated actions between state and federal governments in the allocation of resources, but there must be the same coordination of action connected with the allocation of the environment.

And finally, I mentioned a fourth cause of high cost to industry, governmental actions which benefit other segments of the private sector at the expense of the fishing industry. I will not say a great deal about these because my judgment is that if the root causes which I have been talking about can be dealt with effectively, most of these others will disappear. The trade policy, the anti-trust policy of the United States, the Jones Act which gives unusual benefits to the fishermen injured in connection with their work, and the law against foreign fishing vessels used in the fishing industry are examples. Most of these will probably be substantially relieved by the removal of the root evils, since that will inevitably result in substantial improvement in the cost situation in the United States industry. In the long run, we will be facing a shortage of fishery products rather than an excess, as the demand for fishery products increases in other parts of the world. The anti-trust policy of the United States works against the fishing industry because of the capital structure of the industry. But with the removal of the institutional barriers which I have been talking about, we should see major changes in the capital structure of the industry which would eliminate the disadvantages of the anti-trust policy. The same thing is true in the case of vessels. As the costs drop sharply in the fishing industry the cost of using American vessels will become less and less a disadvantage.

To summarize, the strategy, as I see it, is for government to deal with root problems rather than with symptoms, to address itself to the institutional problems. Because these are extremely delicate and highly controversial and, therefore, extraordinarily difficult to deal with, it is perhaps understandable why government has not dealt with such problems in the past.

But in my judgment this is where the effort must be put and I am happy to say that in the National Marine Fisheries Service this is more and more the pattern of major efforts. It is not without very considerable pain, however, that this is done. These institutional problems are not well understood and because of the political penalties attached to addressing them directly, courses of action which involve their solution are not enormously popular. It is much easier and more painless for the advocates of a strong fishing industry to support the attack on the apparently relatively simple symptoms which we have spent so much time on in the past. This is not the way out, as I see it. It is the responsibility of the government and the leaders in industry and the academic community to insure that attention goes where it should go—to the institutional constraints which make life so difficult for the fisherman.

Needless to say, I am not suggesting that we should ignore short-run problems, some of which fall in the category of what I have called symptoms. I understand that in some cases, patients are prone to death from symptoms, while doctors search for root solutions. The same is true in our business. I am talking about changes in emphasis in the main, with, perhaps, the abandoning of some non–cost-effective activities which we have carried on in the past.

March 1971

11

Fishery Management and the Needs of Developing Countries

J. A. GULLAND

The management of marine fisheries has received considerable attention from those concerned with natural resources public policy for a long time. However, this attention has in the past been almost exclusively focused on the problems of fisheries of the richer developed countries. In the same period bodies such as the Food and Agriculture Organization have been primarily concerned with the problems of developing the fisheries of the poorer countries.

Until quite recently these concerns, with their apparently opposite evaluation of what made up the most pressing problems of the fisheries involved, could co-exist without too much inconsistency. The heavily fished stocks, in need of management, were only those near to the industrial countries. Purely local fisheries, supplying only the local population, offered serious threats only to the most vulnerable stocks. The serious dangers came with the technological advances which allowed the fish to be caught cheaply and supplied to a large market. The steam trawler at sea, and railways and ice-plants on shore resulted in the great reduction of the North Sea stocks of bottom fish such as plaice and cod at the end of the nineteenth century; the Pacific salmon became threatened by the development of the canning industry. By the middle of this century such technological advances had led to management being desirable in a large number of stocks in the North Atlantic and the North Pacific, but still few elsewhere.

J. A. Gulland's early training was as a mathematician. He worked for a number of years at the Fisheries Laboratory, Lowestoft, England, where he specialized on the population dynamics of demersal fish. He has been concerned with providing scientific advice to various international commissions, such as the Northeast Atlantic Fisheries Commission, the International Commission for the Northwest Atlantic Fisheries, and the International Whaling Commission. Since 1966 he has been with the Department of Fisheries, Food and Agriculture Organization of the United Nations, Rome, concerned with fisheries management and planning of fisheries development programs throughout the world.

The views expressed in this paper are those of the author and not necessarily those of the Food and Agriculture Organization.

Since then many stocks in the rest of the world have become heavily exploited. There are two main reasons for this: first, the fishing fleets of the developed countries of the north have moved south, aided by the development of the long range, self-contained freezing trawlers, and of the integrated fleets of fishing and support vessels; second, the poorer countries are developing their own industrial fisheries, especially those countries which have an advancing technology but still have fishermen prepared to work for long hours under difficult conditions for low wages. For example, Korean longline tuna vessels fish all over the world, Ghana has an increasingly efficient fleet of large factory trawlers operating as far afield as Morocco, while Peru has a fish-meal industry capable of catching and processing in three weeks a weight of fish equal to the entire annual United States catch.

These events of the fifties and sixties have changed the nature as well as the scale of the problems of fisheries management. Where more than one country was involved, it could be assumed in the past that the nations concerned could discuss matters on the basis of general similarity of interests. The United States and Canadian halibut fishery, and the International Pacific Halibut Commission, is an obvious example of two countries with very similar interests as regards both the type of industry (vessel, gear, and markets) and also the objectives which should be served by management. Though they have not been explicitly stated, the major objectives are apparently protection of the individual resource and limitation of disturbance to the existing industry. On the other hand developing countries might well, for example, put greater emphasis on maximizing protein protection regardless of individual species.

It could also be assumed that both member countries of the halibut commission were served by reasonably efficient administration, research, and enforcement agencies. The fisherman from the developing countries often feels not only that his fish are being swept from the sea by armadas of huge foreign vessels, but that in any argument about the conservation and management of the resource his side will be outgunned by the weight of scientific, legal, and diplomatic expertise on the side of the rich countries. Though few if any of these thoughts are exclusive to fishermen from developing countries, their justification is greatest in relation to these countries, and represents a real problem in considering management of stocks of fish exploited by both rich and poor countries.

Another problem that is newly arising is that of countries wishing to develop their fisheries in areas or on stocks which are already heavily exploited and subject to a management regime. In the past the potential new entrant was almost certainly not closely connected with the area concerned, and in any case had available to it a choice of other resources which were underexploited and not subject to regulation. While the so-called abstention principle has not been widely accepted, it embodied some very logical arguments, and was until recently not clearly inequitable. So long as there are many alternatives, it was not unreasonable to expect these developing new fisheries to abstain from entering those that were already being fully exploited, and under effective management. However, as the number of underexploited alternatives is reduced, such a principle must increasingly act against such countries as Nigeria, with a large population, but currently only a small and mainly artisanal fishery. Future management schemes need to take into account this desire of the poorer countries to establish their own industrial fisheries as they reach the necessary technical competence.

The present paper discusses the ways in which these interests and concerns of the developing countries—their desires to enter established fisheries, their concern about the effects of long range fleets, and fears of the imbalance between the rich and poor countries—can affect the views, objectives, and techniques of fishery management which have been established in the richer countries.

Resources

The magnitude of the resources in relation to present catches and likely future demand is critical in determining the nature of the management problem. Since World War II the world fish catch has been increasing steadily, doubling every ten to fifteen years. The 1969 marine fish catch was some 56 million tons. Estimates of the potential yield range from figures of under 50 million tons (which limits have already been passed) up to 2,000 million tons (Chapman, 1965).

A lot of this range is accounted for by different definitions of what should be included. The actual production of living matter in the sea is enormous—some 200,000,000,000 tons of plants are produced each year—but virtually all of these, and most of the animals that feed on them, are too small and scattered to be harvested economically.

A recent detailed study by FAO, in collaboration with a large number of scientists and institutions throughout the world (Gulland, 1972), offers estimates of the various living resources of the ocean. For the more familiar types of fish, which can be harvested by well known types of gear—mainly the bottom-living fish such as flounders, sea-breams, and so on, and the shoaling pelagic fish such as herring and anchovies—the estimated potential is only 100 million tons per year. Very large harvests, perhaps of hundreds of millions of tons, could be taken of the smaller and less familiar animals, such as the krill in the Antarctic. The harvesting of these resources will require advanced technology for the catching and marketing, and are therefore of little direct interest to the developing countries. They must concentrate on the more familiar fish, leaving the krill to the Russians and the Japanese, who are, despite their distance from the Antarctic, already carrying out trial fishing there.

A potential of familiar types of fish of around 100 million tons, compared with catches which have been doubling every ten to fifteen years, and which have already passed the 50 million ton mark, suggests that a world-wide crisis of fishery management cannot be far off. There is no reason to believe that the pressures that have increased the number, size, and efficiency of the fishery fleets of the world will not continue, but soon it will be difficult for the surplus capacity exploiting one area or stock to be diverted to an underexploited area or stock. Examples of heavily exploited stocks, with excessive fishing capacity, can now be found in all parts of the world. The fish-meal fishery of Peru has grossly excessive capacity. The shore plants could, if working at full capacity, process the entire world fish catch; the catching side is equally over-expanded, and has to be controlled by two long closed seasons, as well as other restrictions (Instituto del Mar del Peru, 1971 a,b). In the Gulf of Thailand the local fishery was transformed in the early sixties by the introduction of trawling. Catches have risen to half a million tons, but there have been the classic effects of declining catch rates—to less than one half of their original value—and overinvestment in what had been a most successful and profitable fishery. Measures are being taken to restrict the amount of fishing, and to divert those larger vessels that can go so far to areas beyond the northern part of the gulf.

These are fisheries by a single country on a single species, or a group of species that are fished together indiscriminately. The situation off West Africa is much more complex. The richness of the resources, especially in the northern part from Senegal to Morocco, have attracted large fleets from Europe and Asia as well as from the local countries. The state of the stocks has been recently reviewed by a working group set up by the Fishery Committee for the Eastern Central Atlantic (CECAF). Table 12 gives the statistics of total catch in the area as summarized in the report of the group (FAO, 1970a). These fleets exploit a range of stocks, many of which are heavily fished. These latter include the inshore demersal stocks off the Ivory Coast and Nigeria (fished by the local fleets), hake and seabreams (mainly fished by vessels from southern Europe), and cuttlefish (mainly fished by Japan and Spain). The group also expressed concern about the rapidly increasing catches of *Sardinella* by factory ships from South Africa and Norway, as well as from the USSR and local countries.

TABLE 12

CATCHES IN THE EASTERN CENTRAL ATLANTIC, BETWEEN THE CONGO RIVER AND THE STRAITS OF GIBRALTAR
(Thousands of Metric Tons)

Coastal Countries	1966	1967	1968	1969
Angola	0.9	1.0	1.0	1.0
Cameroon	9.0	10.9	12.6	15.5
Cape Verde Islands	4.0	5.9	4.9	4.0
Congo, Dem. Rep. of	12.0	12.4	12.4	12.0
Congo, Pop. Rep. of	11.2	10.6	10.1	9.4
Dahomey	3.8	5.6	5.0	5.0
Equatorial Guinea	1.2	1.0	1.0	1.0
Gabon	2.6	2.6	3.0	3.5
Gambia	3.2	3.4	4.3	4.2
Ghana	74.5	103.1	94.1	140.1
Guinea	4.9	4.9	5.0	5.0
Ivory Coast	57.6	62.9	65.8	67.0
Liberia	11.8	13.5	15.6	18.5
Mauritania	16.0	17.7	18.0	18.0
Morocco	292.9	249.2	207.9	215.6
Nigeria	60.0	66.8	67.0	67.0
Portuguese Guinea	0.7	0.7	1.3	1.7
São Tomé and Principe	0.8	0.9	0.8	0.8
Senegal	116.5	132.0	153.7	162.1
Sierra Leone	31.4	32.7	22.6	24.6
Spanish Sahara	3.8	3.9	3.9	4.0
Togo	4.5	5.0	5.0	5.0
Subtotal	723.3	746.7	715.0	785.0
Noncoastal Countries				
China (Taiwan)	1.5	0.3	6.9	12.0
France	45.8	43.6	57.8	50.5
German Dem. Rep.	0.2	15.5	3.5	3.9
Greece	30.1	31.6	36.8	33.3
Israel	1.5	4.0	3.1	0.9
Italy	64.7	69.4	62.7	45.3
Japan	116.6	170.2	185.2	163.5
Korea, Rep. of	7.1	11.7	12.6	13.6
Norway	0.5	1.2	0.6	2.1
Poland	40.7	44.3	32.9	44.5
Portugal	41.4	39.8	40.0	36.5
Romania	7.1	8.8	5.5	6.0
South Africa	—	—	—	48.0
Spain	181.4	179.4	178.3	178.4
USSR	79.3	153.5	318.6	569.7
United States	—	1.4	10.4	22.5
Subtotal	617.9	759.7	954.9	1,230.7
Total	1,341.2	1,506.4	1,669.9	2,015.7

Present Arrangements

The management of some of these heavily exploited fisheries in developing countries, for example, in Peru or Thailand, involve only one country. The problems raised are much the same as in managing a fishery confined to a single developed country though the balance between various objectives may be different. A developing country may wish to give emphasis to maximizing total catch (rather than the catch of any one individual species), and to minimizing investment costs, particularly those with a foreign exchange element. It may on the other hand be less interested in reducing labor costs, when there is a surplus of unskilled labor. A management problem (using management in a fairly wide sense) that developing countries often face concerns the modernization of a fleet of canoes. Some simple matters, such as some basic training for the fishermen, or the fitting of outboard engines, can perhaps double the catching power of the individual fisherman. If the inshore stocks are underexploited then such action can improve the supply of fish and the income of the individual fisherman. However, it may happen that the stocks available to the canoe fisherman, even with outboard motors, are fully exploited. Increased efficiency will therefore give no extra catch. The conventional answer to this is to reduce the number of fishermen, in proportion to the increased efficiency, for instance having five thousand fishermen with powered canoes, rather than ten thousand with only paddles or sail. This would certainly increase the income of the lucky five thousand, and probably also reduce the price to the consumer. It would, however, only be an acceptable action if alternative work could be found for the surplus five thousand—who will often be unskilled and illiterate.

Multinational fisheries involve the differences between developed and developing countries more distinctly. Two patterns of international bodies concerned with management have so far been developed, typified by the Inter-American Tropical Tuna Commission (IATTC) and the International Commission for the Northwest Atlantic Fisheries (ICNAF), respectively. The former has its own staff for collecting and analyzing data, carrying out background research, and proposing regulations; to do this national governments have to provide the commission with comparatively large contributions. ICNAF has only a small permanent staff, the Northeast Atlantic Fisheries Commission has no permanent staff, and the secretarial work is carried out by the British government on contract; both ICNAF and NEAFC rely on member countries to collect the necessary data, carry out the necessary research, and propose regulations. ICNAF, however, does fulfill a most important research role by providing at its annual meetings, and through special meetings of special scientific working groups, the mechanism by which the various national research programs can be coordinated.

One disadvantage of the present practice, more particularly of the North Atlantic (ICNAF) system, is the long time required to propose and take action compared with the speed of many fishery developments (Gulland, 1971). An important element in reaching any international agreement on management action has generally been that the action concerned is based on sound scientific evidence. Used in the extreme this can be taken to mean that all the scientists concerned should be agreed on the precise effects that would follow from any proposed management action.

Unfortunately fishery biology is not an exact science, and it is very difficult to make precise and accurate assessments of the state of the stocks, and of the effect of regulations. The most accurate methods depend on studying the effect of heavy fishing on the catch per unit effort, or on the age- or size-composition of the stock. These cannot be readily used until there have been some years of heavy fishing. Add to this the time taken to carry out the necessary analyses and hold scientific and administrative discussions to consider the possibilities of introducing regulations. It then becomes clear that this traditional method of operation is very time consuming.

Modern fisheries, on the other hand, can grow extremely rapidly. The time when enough, but not too much, investments of manpower and capital have been made, and when regulation of the amount of fishing can be introduced fairly painlessly, is likely to be reached and passed well before management can be introduced by the traditional pattern. Some speeding up of the scientific process is possible by using lines of approach other than that of classical population dynamics. Thus surveys by research trawlers can give reasonable estimates of the abundance of demersal fish, and hence provide useful guidance of the possible yield, especially of the level of yield at which regulations are likely to be needed (Alverson and Pereyra, 1969).

These difficulties, of having sufficiently early advice to introduce management measures when that can be done comparatively easily, apply to either kind of international commission. They are fewer in the western American (IATTC) type, where a strong head can propose the introduction of regulations earlier, on the basis of less complete analysis, than in the Atlantic type of commission. For both types these difficulties affect both developed and developing countries equally.

The poorer countries have a special concern about the possible inequalities of the two types of commissions. In the western American commissions, such as IATTC or the International Pacific Halibut Commission, the annual dues to the commission can represent a heavy drain on their limited resources, even if the dues are weighted in accordance with the volume or value of the catch. Also these dues have to be paid in foreign currency. Often a poorer country cannot easily afford both to pay commission dues and to engage in its own programme of basic research. It will therefore have to rely entirely on research carried out by the commission, or by richer countries. Because of the probable imbalance in dues paid, and particularly in the availability of suitable people, the staff of the commission is likely to be heavily biased toward the richer countries. Even if deliberate efforts are made to maintain a balance it is not at all clear that the interests of the developing countries are served better by having some of their limited number of good people serving the commission than by staying at home. With the best will in the world the poorer countries are likely to feel that the commission's advice and recommendations will tend to be biased against them. This feeling will act against the capacity of the commission to act early, on the basis of probabilities rather than scientific certainties.

A poorer country's feeling of being the underdog, with scientific and diplomatic skills weighted against it, is likely to be even stronger in the Atlantic type of commission. Some of this suspicion may be unjustified. At least in the North Atlantic the long history of scientific cooperation through the International Council for the Exploration of the Sea (ICES) (founded in 1906), and elsewhere, existing well before the results of the scientific studies would have any direct restrictive effect on national fisheries, has fostered the feeling of international cooperation. Many scientists working in the North Sea would consider themselves almost as much ICES scientists as specifically English, Dutch, or Scottish scientists.

It may, however, be doubted by some whether these independent attitudes of the scientists can be fully maintained when their results will have direct and possibly severe impact on national fisheries. Certainly in some other regions, where there is a shorter period of international scientific collaboration, and where the scientific analyses have been used more immediately in controlling national fisheries, scientists have taken more strictly national attitudes during international meetings.

An intermediate approach between the two types of international commission seems possible, which might be more satisfactory to the poorer or developing countries. In such an arrangement the main input, of scientific research, or in proposing regulation measures, would come from the various member nations, as in the present North Atlantic commissions. However, there would be, in addition, a strong input from the staff of the commission, or from other independent sources. While this input should be essentially technical, for example, in providing additional expertise in the assessment of the state of the fish stocks, it would also help to maintain the balance between the larger or richer countries with strong national capabilities and the smaller or poorer countries which cannot maintain their own expertise in all the relevant fields. For example, at recent meetings of the Research and Statistics Committee of the International Commission for the Conservation of Atlantic Tuna, the input to the scientific discussions of experts from FAO was strongly welcomed by several of the smaller countries. Some of these felt, possibly groundlessly, that there was otherwise a risk of the committee being dominated by the undoubted expertise of the two biggest countries, Japan and the United States.

In this example the independent expertise, or at least some of it, came from outside the regional body concerned. This might be in principle somewhat undesirable, but also somewhat inevitable because FAO, with its much larger staff, is more likely to have individual experts in the various fields concerned. On the other hand the pattern outlined here, of support from FAO (or other independent sources) being merely supplementary to national work, is quite distinct from the more far-reaching schemes in which some international body would have direct responsibility for fishery management throughout the world, which have been occasionally proposed, and perhaps more widely opposed, on a number of grounds.

Objectives of Management

Another big divergence between the interests of established fishing countries and those with developing fisheries concerns the objectives of management. The general objectives of fishery conventions are usually set out in terms of "ensuring the rational utilization of the resource" or, if more specific, of "maintaining the maximum sustainable yield." The practical impact of these objectives is either to preserve the present status, or even, if fishing pressure is very high, to move back to the conditions of some earlier period when the fishing effort had not reached an excessive level. Prevention of change, or still more a reversal of recent changes, is directly opposed to the interests of those wishing to increase and develop their fisheries.

Such conflict is inevitable. If all that can be harvested is already being taken by the existing participants, new entrants can only be accommodated by one of the other countries reducing its share. The extent to which a country might be willing to do this will depend on the persuasiveness of the claim of the new entrant for a share of the fishery. At one extreme few North American fishermen, though perhaps agreeing that Korean fishermen should be able to increase their general activities, would accept that they should be allowed to participate in the already heavily exploited salmon fisheries in the eastern North Pacific. On the other hand, in the tuna fisheries of the Indian Ocean in which the stocks of larger fish (principally yellowfin, bluefin, albacore, and bigeye) are already fully exploited by the long range longline vessels from Japan, Korea, and Taiwan, there would be general agreement that the countries bordering the area should be able to enter into the fishery on these stocks when they are technically capable of so doing.

Where the objectives and regulations are expressed purely in physical or biological terms the need to consider the clash of interests of established and developing countries is not too obvious. Such measures as controls on the sizes of fish caught, limitation on the mesh sizes of nets, and also closed areas or closed seasons, when they are applied purely for biological reasons, for example, to protect nursery areas, should reasonably be expected to be adopted by all new entrants.

Simple direct controls on the total catch, or total amount of fishing, imposed as a single quota, with all fishing stopping when the quota is reached, also appear to be nondiscriminatory and might then be expected to be accepted by potential new entrants. The lack of discrimination is perhaps illusory. For example, long intergovernmental discussions have been held concerning the possible differential effects of the regulations of the Inter-American Tropical Tuna Commission. A quota for the yellowfin tuna has been in operation for a few years, with the usual result of a steadily shortening season. A short season adds to the costs of all participants, but probably hits particularly hard the new entrants with less experience of the fishery. Also the vessels from developing countries, such as Mexico, are likely to be smaller than those from the United States, and are less able to move out of the regulated area than the larger purse-seiners. Many of these have reduced the impact of the closed season in the eastern Pacific by moving into the tropical Atlantic in the second half of the year. Because of these differences and inequalities the Latin American members of IATTC have successfully claimed rights for some vessels to continue fishing after the end of the main open season.

Control of the amount of fishing based purely on biological considerations, without taking into account economic factors, is being increasingly strongly criticized. Some of the biggest gains to be obtained from management are the reduction in costs that could accompany reduction in fishing mortality. All of these are likely to be lost, through some form of enforced inefficiency, unless the regulations are framed to take economic factors into account. Thus the use of a single unallocated quota has resulted in a steadily shortening season and correspondingly increasing costs in such diverse fisheries as Antarctic whaling, halibut, and the yellowfin tuna in the eastern Pacific.

Of the multinational fishery commissions the International Commission for the Northwest Atlantic Fisheries has given most detailed consideration to these general problems, especially through its working group on joint biological and economic assessment of conservation actions (ICNAF, 1968). This group concluded that where the social and economic structures of the countries varied widely, as they do in ICNAF, the most useful management technique was the allocation of quotas to countries. This would allow each country to adjust its own national fishery to make the best use of its quota in accordance with its national objectives (maximum net economic yield, high employment, etc.). For any developing country which is already taking part in the fishery, and does not wish to increase its participation to any significant extent, this is a perfectly acceptable scheme.

Any application of this scheme immediately raises the problem of how the national quotas should be allocated. This has been discussed in principle by ICNAF and others, particularly in ICNAF's Standing Committee on Regulatory Measures. This committee has concluded that while part of the allocation should be made on the basis of historic performance, that is, in proportion to the average catches over some preceding period, another part of the allocation should be made according to special needs. These needs could include those of coastal states, or countries with developing fisheries. Though ICNAF member-states are confined to North America and Europe (plus very recently Japan), they include countries such as Poland that are in the process of developing their fisheries. These have thus some interests in common with those of the third world of Asia, Africa, and Latin America. However, the Polish fishing fleet has larger, more powerful, and in many respects more technologically advanced vessels than those of many longer established fleets in the ICNAF area.

In any case taking account at the time of an initial allocation of the desires at that time, for further development, does not help those countries wishing to develop their fisheries at some later date. As Crutchfield (1970) states, this is the problem of limited entry, usually discussed within the context of one country, transferred to the international field. Just as nationally it is economically advantageous to limit entry by prohibiting new entrants to a fully exploited fishery (unless they arrange to take over an existing share), so, Crutchfield suggests, fully exploited and regulated international fisheries should be closed to new national entrants.

One exception is noted by Crutchfield. African countries should be entitled to enter the trawl fisheries off their coasts which are being heavily exploited by long-range vessels from eastern Europe and elsewhere. This is a clearly acceptable proposition, but it is not so clear that it provides a complete solution even to the problems of the West African countries. There are two reasons why the African countries should be allowed to increase their share of the catches: they are adjacent to the resources and they need the fish. These reasons do not apply equally or in the same proportions to all African countries. Nigeria for instance has a very large population, and is a large importer of fish, particularly of frozen fish taken by European trawlers off West Africa, as well as dried cod from the North Atlantic. The fish resources off Nigeria are limited, and though these imports could be replaced by fish caught by Nigerian vessels off West Africa, much of these catches would have to be taken in waters well to the north of Nigeria. Mauritania, on the other hand, has rich fish resources off its coast, but only a small population, and no immediate prospect of wishing to harvest all these resources itself, at least for internal consumption. If African countries could arrange affairs among themselves amicably, so that there would be, for example, no dispute between Nigeria and Mauritania about their shares of the "African" part of any overall quota, these differences might not matter. Experience elsewhere suggests that this is unlikely. Fishermen from two countries in the same region, for example, the United States and Canada, can readily agree about the wicked depredations of fishermen from outside the area, such as Russians, but this makes them no less willing to complain about each other's activities. All South American fishermen may approve the arrest of United States tuna seiners, but a Chilean purse-seiner or Ecuadorian trawler can still be equally quickly arrested if they fish in Peruvian waters.

The exception proposed for West Africa, which might be extended on the same combination of local interests and need for fish to cover, for example, the development of tuna fisheries in the Indian Ocean by, for example, India or Tanzania, would still offer little to a country which might wish to develop fisheries outside its local region. So long as Nigeria imports substantial quantities of cod it is not unreasonable for it to hope to catch this cod with its own vessels. Its claim to a share of, say, the cod in the Barents Sea is weak compared with those of Norway or the USSR, or even perhaps the United Kingdom or Germany, but only historic interest can give, for example, Spain or Poland much greater rights to the cod on the Grand Banks of Newfoundland.

Up till the present the problem of new entrants has received little attention from international fishing bodies. This is not very surprising, since those few bodies which have introduced systems of limiting the amount of fishing have been naturally in areas of very intense fishing. The pattern of change has been for effort to leave these areas, rather than to enter it from other regions. The only multinational fishery with agreed allocation of catch quotas is that for Antarctic whales. This agreement comes to an end if there are new entrants, though the possibility of this is slight owing to the great capital cost of a pelagic whaling expedition.

A body that has discussed this problem is the FAO's Indian Ocean Fishery Commission. Its committee on Management of Indian Ocean Tuna considered the problem in relation to the stocks of large tuna, which are heavily exploited, almost entirely by countries from outside the region. It drew up some general principles, which are worth quoting in full (FAO, 1970b).

1. The stocks must be maintained at a level which can provide a high sustained yield.
2. Action to conserve one stock should not interfere with the development of fisheries on other stocks which are still under-exploited.
3. Management measures should be so framed that, whilst conserving the resources, they would afford the opportunity to countries not yet significantly participating in Indian Ocean tuna fisheries to build up their fishery industry within a reasonable period to associate themselves effectively with programmes of rational utilization on a basis of equality.

These are of course only very general guidelines, and are open to a variety of interpretations. What for instance does "equality" mean? If it means that each country bordering the Indian Ocean should be able to catch as much large tuna as those already fishing, then the prospects of effective tuna management are dim. The main values of these guidelines are first, a recognition of the problem, and second, acknowledgment by those already fishing that special arrangements are needed for developing countries.

I have here discussed the problem mainly from the point of view of the developing countries, wishing to expand their fisheries. Those already exploiting the fishery, and especially those that have made great efforts (often including substantial short term sacrifices) to manage and regulate these fisheries, may have quite different opinions. However, it is generally agreed that the richer countries have a special responsibility for assisting the poor countries to develop, and in this bodies such as FAO have an important role. This development should include the development of fisheries, and it may be that in the future one way that assistance can be given will be by granting developing countries especially favorable terms for participation in management schemes. Internal political difficulties in the richer countries might be avoided or reduced by using aid funds for compensating the fishermen concerned.

Greater attention to the problem of the developing countries in relation to international fisheries management is also of direct concern to the established countries. Management of high seas fisheries requires the agreement and participation of all those engaged in the fishery. If the developing countries feel that they can only participate under disadvantages, they may refuse to participate. Coastal countries may attempt to protect their interests by claims to wide jurisdiction on fishery matters. Other countries may decide to fish without regard to management measures, introduced without what they may feel to be sufficient attention to their interests. These latter could include countries of the size, and potential demand for fish, of China and India (neither of which has so far gone in for more than coastal fishing). Management schemes can hardly be viable if these large countries refuse to participate.

February 1971

REFERENCES

Alverson, D. L., and W. T. Pereyra. 1969. "Demersal Fish Explorations in the Northeastern Pacific Ocean." *Journal of the Fisheries Research Board of Canada,* 26 (No. 8): 1985–2001.

Chapman, W. M. 1965. *Potential Resources of the Ocean.* Long Beach, Cal.: Van Camp Sea Food. 43 pp.

Crutchfield, J. A. 1970. "Economic Aspects of International Fishing Conventions." In A. D. Scott (ed.), *Economics of Fisheries Management,* pp. 63–77. H. R. Macmillan Lectures in Fisheries. Vancouver: University of British Columbia, Institute of Animal Resource Ecology.

Food and Agriculture Organization. 1907a. *Report of the CECAF Working Party on Regulatory Measures for Demersal Stocks.* FAO Fisheries Reports, No. 91.

———. 1970b. *Report of the Second Session of the Indian Ocean Fishery Commission.* FAO Fisheries Reports, No. 95.

Gulland, J. A. 1971. "Science and Fishery Management." *Journal du Conseil,* Vol. 33, No. 3.

———. 1972. *The Fish Resources of the Ocean.* London: Fishing News (Books) Ltd. 255 pp.

Instituto del Mar del Peru. 1971a. "Report of the Panel of Experts on Population Dynamics of Peruvian Anchoveta." *Boletin del Instituto del Mar del Peru,* 2 (No. 6): 324–72.

———. 1971b. *Report of the Experts Panel on the Economic Effects of Alternative Regulatory Measures in the Peruvian Anchoveta Fishery.* Informe Instituto del Mar del Peru, No. 34. 83 pp.

12

A Confidential Memorandum on Fisheries Science

PETER A. LARKIN

In the expansive mood of the nineteenth century it was believed that the ocean was an unlimited producer of fish for man's use. This mood was replaced in the early part of the twentieth century by the realization that fish populations could be depleted, and so should be harvested to produce a maximum sustainable yield. In consequence, for many years now, it has been conventional to speak of maximum sustained yield as a prime objective of fisheries management. For several decades fisheries biologists have been preoccupied with ever more precise methods for estimating maximum sustained yield, given various restraints. There is now a substantial literature which provides what appears to be a good basis for going a long way toward the declared objective of obtaining maximum sustained yield. For example, knowing parameters of growth and mortal-it occasioned by natural factors and by fishing, and the relation between the size of a spawning stock and the number of recruits it produces, we can, for any stated mesh size, specify the rate of fishing that will produce maximum yield (or, for any rate of fishing, specify the appropriate mesh size).

Just as we were gaining the technical skills to manage for sustained biological yield, there was growing realization that yield of dollars might be more important. The value of the catch in relation to economic effort may be more germane to human predation than catch in relation to physical effort. Again, a fair sized literature of fisheries-oriented economics is now available to guide us to what seems to be rational economic use of fisheries resources.

Peter A. Larkin is head of the Department of Zoology, University of British Columbia. He was a Rhodes Scholar. He has held positions as director, Institute of Fisheries, University of British Columbia, and director of the Fisheries Research Board of Canada Biological Station, Nanaimo, British Columbia. He has served in several positions on the Science Council of Canada and is a member of the Special Committee on Problems of the Environment. Dr. Larkin has published widely in the field of fisheries biology with specific reference to some of the salmonid fisheries of Canada.

Of course, these are not the kind of generalizations that appeal much to the perfectionists in the fraternity of fisheries biologists and, admittedly, there is still lots more to be known. For example, for many of our fisheries the relation between stock and recruitment "remains obscure," by which we mean that it is the same relation that one would observe if there were no relation. In other instances it is difficult to estimate fishing effort because of rapidly changing fishing technologies. The consequences of harvesting mixed species continue to haunt us like a can of many kinds of worms. Even on relatively basic matters, such as the genetic consequences of harvesting, we are much in the dark.

Nevertheless, for all these shortcomings, it seems fair to say that by now we are able (or should be able) in at least some places, for some species of fish, to manage for maximum sustained yield. We have had fifty years of research, have accumulated an impressive literature, and certainly can convey to undergraduates an impressive savoir faire. However, if we try to write a book on the *accomplishments* of fisheries management toward the objective of maximum sustained yield, I think we will conclude that we need twenty pages for introduction, one page for results, and more than one hundred pages for rationalized excuses. In brief, for all of our knowledge, we have not frequently demonstrated an ability for management.

Considering first the freshwater fisheries, we have, for example, the story of the fisheries on the Great Lakes, a bewildering tangle of regional regulations, introductions of exotic species, and environmental effects, which is of no value as an example of how to manage, even though the Great Lakes have long been adjacent to a strong concentration of fisheries biologists. As a matter of fact, I know of no lake in Canada which is currently managed for maximum sustained yield for either a commercial or sport fishery. There may well be one, or even five, but there are certainly very few. Much the same is true, I suspect, of the United States.

If United States management practices for freshwater are like those in Canada, then one may observe:

1. Sport fisheries on "accessible" lakes and streams (i.e., within one hundred yards of a road) are often characterized by overexploitation, and are subsidized with hatchery plantings. Sport fisheries on inaccessible lakes and streams are trivial. For virtually all sport fisheries there is no restriction on effort and only arbitrary regulation of kinds of gear. Such regulation as there is has no logical connection to concepts of maximum sustained yield.

2. Commercial fisheries on lakes may also overexploit stocks in areas close to markets, but underexploit stocks in remote areas. Most of Canada's lakes, for example, are small, remote, and unproductive. Man-

aging their fisheries for sustained yield poses major problems of investigation, logistics, and economics.

3. The only freshwater fisheries that *are* characterized by some notions of maximum sustained yield are fish farming enterprises of one kind or another, but their operations are not guided by the usual considerations of fisheries management.

For marine ground and pelagic fisheries, we can make a circumstantial case that there has been maximum sustained yield management for some fisheries; for example, halibut and, for a few years before the fishery collapsed, herring of the British Columbia coast. The same is true of the much richer fisheries of the east coast of North America, for which international competition has been a major complicating factor in attempts to aim at maximum sustained yield. Our performance is not bettered by our colleagues abroad. I know of no European, or Asian, or African, or South American fishery which is clearly being managed for maximum sustained yield, with the possible exceptions of the Peruvian anchovy fishery and the yellowfin tuna fishery of the eastern tropical Pacific.

When it comes to management for maximum economic yield, it is very doubtful that any marine or freshwater fishery in the world can be pointed out as a working example. Perhaps in the long term strategies of economic competition there is some sense to be made of the typical patterns of overcapitalization, but certainly in the restricted context of fisheries we can only conclude that man's global harvest of seventy million tons a year is taken in an awesomely expensive and inefficient manner.

Our local west coast salmon fisheries are often cited as fine examples of management for maximum sustained yield, but on close examination they are more reminiscent of the consequences of a long war of attrition. In the aggregate, the catch of salmon on both sides of the Pacific is substantially less than it was in the 1920s, despite the fact that we now know far, far more about the various species and populations. To some degree this has been a consequence of harmful effects on the environment of salmon, only partially offset by protection measures, and efforts (one hundred years of efforts!) to augment natural production by hatcheries and related devices. But these considerations aside, it is still patently obvious that we have not been able to manage for optimum escapement, which is, for salmon, *the* requirement for maximum sustained yield.

In my view, since the 1920s we have been overly impressed with the homing of salmon, forgetting that (1) salmon do stray from one stream to another; (2) most so-called stocks are mixtures of substocks; and (3) many of our fisheries harvest mixed stocks, mixed species, and salmon of mixed national origin. In these circumstances, it is virtually impossible to manage according to the precepts of maximum sustained yield without making compromises. Even two stocks with identical population parameters, but influenced by differing environmental factors, cannot be managed to give their separate maximum sustained yields when they are harvested jointly!

When one adds in the economics, the present practices of our salmon fisheries border on farce. We have far more fishing capability than we need, and it is divided up into small, inefficient units. There are only limited concepts of fleet strategy. Our regulations seem to be devised to limit simultaneously the catch and the efficiency with which it is taken. To economists, our salmon fisheries are a notorious example of the perils of permissiveness.

There is nothing new in all of the foregoing. We all know it, and until recently, we have all imagined that better days would come. Given time to discover and educate, we felt we could eventually recover and legislate. But about now many of us are beginning to wonder if these happy days will ever arrive, if we persist in our present practices of research and management.

We must first acknowledge that, for the most part, our theories of fisheries management are essentially based on circumstantial evidence. To the degree that I can look at the fisheries literature with detachment, I would say that its striking characteristic is its failure to test its hypotheses with experiment. It is one of the standard chestnuts of undergraduate teaching to describe the amateur's notion of what constitutes a scientific experiment. The amateur invariably says, "Wouldn't it be interesting if we tried this?" After "this" has been tried, one of course never knows what would have happened if "this" had not been tried. We therefore stress to the undergraduate the importance of doing experiments with controls. Very little of our fisheries management is conducted in this experimental tradition. Most commonly, we attempt to weave from the evidence what seems to be a credible account of what happened, and in consequence, we have some fine controversies that will be forever unresolved because we cannot go back and do it all another way. The California sardines are a fine example—they are virtually gone now, but the arguments persist on whether to blame the fishery or the change in ocean conditions (or both, or neither).

Admittedly, opportunities for experiment are difficult to contrive in natural conditions, especially when there are livelihoods at stake, but experiment we must if fisheries investigation is ever to graduate into the realm of science.

The only alternative to experiment as the method of science is the use of provisional hypotheses which are clearly stated and which make predictions. If the prediction is correct, the hypothesis is vindicated and retained; when it is incorrect, the hypothesis is revised and then used as a basis for the next prediction. By scrupulously honest trial and error, one slowly gains knowledge.

Contrast this with the usual vague and fluffy projections of many fisheries managers. With a prolific use of adjectives and adverbs, conditional sentence structure, ambiguities, and/or characteristic sins of omission, we usually play a game of "making predictions" which can be constructed as being correct, no matter what happens. In a very real sense, we do ourselves the disservice of failing to take advantage of the opportunity to learn by clearly stating our ignorance. We then further confound ourselves by failing to clearly document why certain management decisions were made. In consequence, when we view our performance in retrospect, we can conveniently forget the confusion at the time of the events. We can rationalize whatever we did as consistent with whatever we predicted, and with whatever are the vaguely stated wisdoms of our witchcraft. Honestly written, many of our fisheries sagas would be like mysteries in which several culprits are simultaneously and unwittingly involved in murdering each other and the reader. In brief, our fisheries literature is largely unscientific in the strict sense of the word, and our fisheries management is unscientific in almost any sense of the word. We are a long way from having the experimental, or empirical, base on which good science relies.

This state of affairs perhaps stems in part from the fact that fisheries biologists may not be, on the average, as bright as other scientists, and are, perhaps, conditioned at a relatively early age not to be too scientific. But I refuse to believe that this is the whole explanation. First, I think we have been plagued with the difficulties of coping with a very rapidly changing world. Second, in many instances the economic returns from the fisheries concerned would not justify the kind of expenditure that would be necessary to put the fisheries on a scientific basis. But much the most serious cause for the present state of affairs is that most of the decisions of management are made by commissions or by administrators for whom the biological information "is only one of the many factors to be considered." It is no doubt wise to make decisions on this kind of enlightened broad basis, but it may also be a euphemistic way of camouflaging that the people charged with making the decisions do not understand what science is or how it progresses. I do not mean this in a derogatory way; I just state it as a fact.

Let's examine some of the decision-making machinery in fisheries. Characteristically, decision making in fisheries management involves, either directly or indirectly, consultation with representatives of all the groups that may be affected—industry, fishermen, experts, ethnic groups, social groups, and, in a vague sort of way, the people at large. In these circumstances, the decisions are bound to have a certain character.

"Why rock the boat?" is the usual theme of representatives. If disaster has not struck in the last three or four years, if things are more or less the same, or only a little bit worse, why invite chaos or catastrophe? If things go wrong, even though we do the same as we did before, then we can surely say that what happened must have been God's fault, not ours. And so we proceed, step by ignorant step, rarely experimenting, always rationalizing the compromises, never making original mistakes, but always in a position to be technically blameless, following a stumbling way which leads to anxiety and concern without the relief of inspiration and new knowledge. In brief, we have contrived a scheme for managing fisheries which substantially impedes the rate of gaining new understanding. In a word, it is amateurish.

I would now like, with an example, to illustrate my notion of what kind of decision making would bring us to an evaluation of what we know, and lead us to better levels of management. To avoid hurting any sensitive feelings, we will deal with an imaginary stock of fish that is managed by an informed, honest, and fearless committee (also imaginary). The fishing season is still six months off and the committee makes its first press release.

THIS YEAR'S RUN COULD BE ALMOST ANYTHING, SAYS COMMITTEE

The Committee for management of the fish fishery announced today that they had reviewed the evidence available to them, applied all of the current theories of management, and decided that they didn't have much confidence in predicting what would happen in this year's fishery. The present techniques of analysis enable three estimates which differ widely; averaging the three estimates, which is probably not warranted, gives a predicted run of 1,700,000 with 95 percent confidence limits of 200,000 and 28,000,000. This is such a wide spread that this year the fishery will be run so as to yield as much information as possible without destroying the stock, creating financial ruin, or causing extreme social stress. The Committee will manage the fishery on the following basis. In Area A, the Ricker stock-recruit curve will be used as the base. Fish in excess of 300,000 should thus be caught and this will be attempted by manipulating the hours of closure according to the following schedule. . . .

And so on in like fashion. The important thing is that the committee states clearly and in as much detail as possible, "This is what we plan to do, and why."

The committee must then rigidly adhere to its stated plan. The only intraseason adjustments that are allowed are those that were anticipated by exact rules in the first press release. This is the way that science progresses—painful realization of ignorance by brutally honest self-assessment. Admittedly, every year over a period of several years the second press release (after the season) might read

FISH COMMITTEE DISMISSED FOR INCOMPETENCE

But, if it is openly acknowledged that only the utmost candor will lead to progress in fisheries science, as in any science, one would hope that the public would accept the necessary mistakes as part of the price. By following this type of pattern we would get a better idea of what we know and which of our theories are correct. We would have exactly the image we deserve, and it might well be better than the one we now have. Most important, we would stop paying lip service to puritanical notions like maximum sustained yield and come to grips with the more complex problems of real life resource management.

This third point bears enlarging because it perhaps suggests the criteria that might be used for deciding on management in our Utopia of scientific honesty. In the first place, we should feel no compulsion to take maximum sustained yield just because it is there to take. After all, every year we make no use of a potential crop of robins or sea urchins, so why should we feel badly if we do not harvest the deer or the salmon? The only possible reason for feeling that it is a duty to harvest is if future harvests of the same or even larger size are thereby ensured—that is, that by being more numerous, the animals can do harm to our future prospects. The scientific evidence is far from showing the severe density dependent mechanism that this view requires; but, even if the evidence were unequivocal, it would still mean that we could choose between harvesting a lot from a stable population, or harvesting less from cyclically fluctuating populations. Maximum sustained yield is only our duty if we choose to make it so.

Moreover, we should not hesitate to take more than the maximum sustained yield if the economic spirit moves us. Given certain, not unreasonable, economic assumptions, it may be desirable to periodically "overexploit" populations of slow growing and long lived fishes, rather than to try to keep going an uneconomic maximum sustained yield fishery. To be most controversial about it, isn't it worth asking the question, "Should we fish halibut almost to extinction in the next five years and then close the fishery for twenty years?"

But just as yield of flesh is arbitrary, so is maximum dollar yield, for it is just as much a technical judgment as maximum sustained yield. We should not feel compelled to be economically efficient; but we should be economically inefficient to the degree we choose.

The real question, then, is what to choose as a level of economy inefficiency, and this is quite clearly a social decision. Enlightened resource management then becomes the application of our expert knowledge to achievement of what is perceived as the social objective, not the maximization of yield, or dollars, or of anything else that is conveniently definable.

The problem of clearly defining social objectives is very much with us nowadays. Until everyone agrees, and agrees to stay agreed, it would not be possible to generate a completely comprehensive set of social objectives. Additionally, it is quite clear that complete agreement would represent the end of social evolution. In these circumstances it is not desirable to be rigid, and what we need is a shifting set of social objectives which change with the times. This is, of course, the politician's bailiwick. It is his job to guess what most people will think is best for them.

We are now full circle and seem to be back to advocating the very type of representative consultation that we now have, and which has proven to be so ineffectual: but there is an important difference. Rather than pretending to be wise, we would acknowledge our ignorance, and boldly *experiment* with new ways of managing resources for the public benefit.

Then the technical question becomes what would happen "if." This I see as the great role of simulation modeling. Its object should be to explore the widest spectrum of consequences of the widest range of alternative policies. The present development of this kind of approach is heartening, because it bears the promise of allowing the representatives, who manage, the opportunity to gauge scientifically the ways of compromising various possible objectives. The aim of fisheries science becomes maximization of understanding so that whatever the undefinable and shifting social objectives, there will be some notion of how to get there. Our real obligation to the future is to let others know how it all works, leaving them enough options to do what they wish in the future.

In this context I see two shortcomings in present modeling work. First is the danger that the models will be believed. We all know that they should not be, but our customers, the decision-makers, like most good salesmen, are notoriously gullible. Their audience, on the average, is overly impressed with hardware that has red and green lights, and software that has cryptic holes. It is a modern substitute for a witch doctor's mask. Our challenge, then, is to sell our product on the difficult sales pitch that it is needed but wrong. The best gambit seems to be that it should be thought of as an extension to the powers of imagination.

This leads to the second shortcoming I see. Most models that are currently being developed are pooh-poohed by traditionalists as pie-in-the-sky. It is a much more penetrating criticism to say that most model schemes are painfully unimaginative, largely dealing with exhaustive but trivial extrapolations of a set of assumptions that are already passé. "Future shock" there may be, but as yet it does not have much real voltage. So far it is a technical future shock. The philosophical impact can be, should be, and eventually will be, much more profound.

In the course of these papers it has been indicated that performance in other public management sectors is just as weak as it is in fisheries. This is perhaps true, and underlines that present patterns of renewable resource use are almost universally unscientific, are not making full use of technical knowledge, and are not capitalizing on our capacities for learning by experiment. Hopefully, in its traditional role as the modest leader in the resource sciences, fisheries will lead the way to new patterns of management and new understanding.

May 1971

13

The Need for Analysis in the Development of United States Fisheries Policy

BRIAN J. ROTHSCHILD

This paper discusses aspects of a methodology for generating public policy on the allocation of fishery resources to the fisheries, or to other users of these resources. At the outset, it is important to note that policies which have, or may have, substantive effects on fishery resource allocation do not all emanate from commensurably high-level decision centers —centers which not only are at a high level in the decision hierarchy, but which also conventionally base their decisions on analyses that are on at least as high a level as the kinds of decisions that are being made.

Because fishery policies which have substantive effects upon allocation are made at various levels in the decision hierarchy (e.g., Department of Commerce, regional office, laboratory, individual) and in various branches of the government (e.g., Department of Commerce, Department of Interior, Department of Health, Education, and Welfare, as well as in the several state governments), it is easy to see that policies at lower levels may not be consonant with those made at higher levels and that the policies made at some levels, in some branches of government, might, when juxtaposed with policies made at other levels or in different branches of government, actually be counterproductive.

The generation of fishery policy from the interstices of this complex jurisdictional web makes it difficult to appraise and evaluate the system

Brian J. Rothschild is director of the Southwest Fisheries Center, National Marine Fisheries Service, National Oceanic and Atmospheric Administration. He was formerly deputy director of the Northwest Fisheries Center and, at the time this paper was written, professor of fisheries at the University of Washington. Dr. Rothchild has been associated with the Honolulu Biological Laboratory of the Bureau of Commercial Fisheries, the Maine Department of Inland Fisheries and Game, and the New Jersey Division of Fish and Game. He has also served as a consultant to various fisheries programs, such as the United Nations Development Program/Special Fund project in Korea and the Indian Ocean Fisheries Program.

that generates the policy This difficulty has promoted a view, among many members of the fishing industry, that the United States does not, in fact, have a fishery policy. This interpretation is, however, not correct. The United States does have a policy. This policy is reflected in a consolidation of the decisions that are made in the various branches of government at the different hierarchial levels. It is, however, true that this policy has not been clearly elucidated. Furthermore it is quite significant that the policy is a conglomeration of decisions that have been made on a more or less *ad hoc* basis, whereas it would be much more desirable to have decisions that arise from the fundamentally sound policy.

If we are to have the opportunity to make the best use of our fishery resources it is essential that we have a unified, carefully constructed fishery policy. Such a policy is a needed guideline for effective decision making. Perhaps, more importantly, a stated policy provides a substrate for debate and a mechanism without which improvement of the extant policy is at the very least inefficient, if not logically and practically impossible. Without a total policy we may still acknowledge that the various and sundry activities that we pursue in attempting to improve the ways in which we utilize our fishery resources are, for the most part, all worthwhile. This conclusion, however, provides little guidance—it leads to organizations that are engaged in activities rather than organizations that are charged with producing specific results. Everything cannot be equally worthwhile. As long as we have finite budgets our scope of action will be limited, and we must choose very carefully among our list of opportunities. Basically we cannot evaluate activities until we talk about specific objectives, programs to accomplish these objectives, and the costs of these programs. Only in this context can we evaluate activities in terms of the financial resources that can be allocated to each. It is clear that in this context many activities—including activities in which we may already have considerable investment—are simply not going to be worth doing.

The consequences of our present approach to fisheries policy are of course manifold. Dominant among these is a temporal trend in which United States boats catch relatively fewer fish while imports of fish and fish products increase substantially. Thus, the segment of the industry that handles fish after the fish are unloaded from the fishing vessel is tending to grow, whereas the segment of industry that strives to place the fish on the dock is not growing. Is this situation good or is it bad? Should it be modified, and if it should, what are the best ways of achieving the modification? In my opinion we have no general answers to these questions. I am not sure—without even approximate valuation of our fishery resources, and the valuation of opportunities which are either afforded or foregone in the pursuit of these resources—that it is satisfactory to say, looking into the future, that it is, for example, economically sensible for the United States flag fleet to be relatively stagnant. In terms of the *status quo*, the economic infeasibility of a growing United States flag fleet may be testified to by the open-access problem and the variety of institutional constraints which are imposed on our fisheries by our extant fishery policies. We cannot, however, simply think in terms of the *status quo*. We need to look into the future and analyze alternative ways of achieving our objectives.

In general, this analysis is needed because in fisheries policy, decisions are continually and typically being made without the decision-makers having other than intuitive ideas about the consequences of their decisions. Yet problems tend to be attacked with the same methodology that was useful several decades ago when the problems were, in fact, much simpler. It is necessary to assert that consistently good decisions can only arise from an understanding of their consequences, and such an understanding, in the complex situations that face us in contemporary problems, can only be arrived at through analysis which is explicitly designed to aid decision making. In resources as large as fisheries we cannot afford to continue to make decisions on an *ad hoc* basis, from a position of pressure, rather than making decisions based upon a well-founded policy. We need analysis to determine when decisions will have to be made and to assemble on a timely basis the information requisite for this analysis.

The fact that appropriate analytical techniques have not been employed leaves us with a dearth of useful information on the following crucial problems or situations:

Poorly developed criteria for national and international temporal and spatial allocation of stocks

The open-access problem

Many overfished stocks

Little understanding of the stock and recruitment problem

Poorly developed theory on multiple species fisheries and the effects of exploitation of one species on the exploitation of others

Fleets capable of exerting tremendous amounts of fishing intensity

Misallocation of stock complexes in the time stream

A need for the fishing community as a whole to participate in rational management and to be held accountable for mismanagement, overfishing, and irrelevant research

A need to improve our understanding of the ways in which fisheries can contribute to the economic development of the developing nations

A limited understanding of ocean-fish interaction as testified to by the limited success in predicting fish abundance and distribution from environmental data

A need to ensure planning for fishery development and management, particularly in the developing countries, vis-à-vis United Nations Development Program funding

A need to develop both biological and economic theory for treating short-term fluctuations. Most available theory depends upon "on-the-average" behavior. Incentives for more efficient stock utlization require predictions of what will happen "next year."

The lack of efficient information systems for the storage and retrieval of catch, effort, and other fishery data

Many ineffective fishery management procedures and concepts

A need to develop theoretical and rational bases for jurisdiction

A consideration of this list raises two particularly important policy-related problems, solutions to which must be approached with whatever urgency can be mustered. The first problem involves the legal regime which governs maritime affairs and the second involves the development of mechanisms by which the United States can more effectively manage its fishery resources. With respect to the former, there are three dominant features. The first is that the present legal regime is very dynamic. Several states have extended their jurisdiction and the more or less traditional order, in addition to these modifications, is continually further being modified or eroded by frequent challenges to the *status quo* when, for example, the fishing boat of one nation fishes in waters or on a stock claimed by another nation. The second is the possibility of a Law of the Sea Conference in 1973. There is a tendency to place LOS 73 at the forefront of consideration in terms of the legal regime, but it is not certain that the conference will even be held. The relative importance of LOS 73 needs to be assessed relative to the everyday dynamics of the legal regime. In any event it seems, based on some of the precedents, that a jurisdictional boundary of perhaps two hundred miles will be a popular proposal. Again analysis is needed. What is the logic behind the magnitude of any jurisdictional boundary? Where did it arise? A fixed boundary can only be good "on the average." It is certainly better for some nations than for others. Such a boundary may not even be good on the average. Again the problem has not been submitted to analysis. From a United States point of view, at least, any existing analysis is not public information, perhaps for very good reasons. But it may be useful to know that such an analysis exists and the criteria that are being employed in judging various alternatives. The third feature, to which we alluded in discussing the second, is that the construction of a modified legal regime does not appear to be based an analyses by a sufficient spectrum of experts and professionals in which objectives (even if they are conflicting) are set forth, alternative ways of approaching the objectives are identified, and the criteria for evaluating the alternative approaches are established. Again, only with appropriate analysis can we understand the consequences of our decisions; if we do not appreciate the consequences, then there is constraint against the effectiveness of the decisions.

With respect to the second problem we can see an important constraint in the development of a United States management rationale. This is that the administrative function of managing our fisheries is enjoyed primarily by the coastal United States. The jurisdiction of the individual states in most cases does not, and most likely would not, cover the entire range of each individual stock. The budgetary resources of each individual state would not be sufficient to attain a minimal research and management program for many stocks and the conduct of such programs in several states would clearly be duplicative. The management of the fish stocks is obviously an interstate problem yet, as mentioned previously, the federal government can exert only limited jurisdiction in this area. There have been proposals to extend the already existing relations between the states and the federal government, but, again, it is apparent that arrangements which have been discussed so far do not provide clear lines of responsibility for particular resources and that alternatives such as *effective* coastal commissions have not been examined. Again analysis is needed.

The reader who is familiar with systems analysis will have known that my purposely tedious use of the word "analysis" refers not simply to the need for additional research but rather to a systematic evaluation of the objectives, alternatives, and criteria used in developing fishery policy. I should like to comment in the remaining portions of this paper upon some aspects of this systems methodology which is an introduction to the procedures and to define more carefully the term "analysis" that was used in the preceding portion.

The discipline of systems analysis has its roots in the development of operations research (Morse, 1970) and is a methodology for providing advice to decision-makers on complex problems which usually involve policy. It is, therefore, no accident that the continued development of systems analysis appears to parallel the continually increasing urgency with which we need to solve the complex problems of our contemporaneous world. In these complex problems the systems analyst endeavors to use systems methodology to ask the "right questions."

In our discussion of systems analysis, it is important to recognize that what we are describing is a particular methodology. It is not necessarily related, for example, to control-system theory, to computer systems, or to the study of ecosystems. Its orientation is considerably different from other methodologies, such as the scientific method or operations research. The scientific method, for example, has always been philosophically oriented toward seeking, but never attaining, ultimate truth (see Copi,

1953) whereas operations research is basically a collection of mathematical techniques (such as mathematical programing, simulation, inventory theory, and queuing theory) which are useful for the solution of relatively well-defined problems. Because the sponsor of the systems analysis needs advice to make timely decisions he frequently cannot afford the luxury of a scientific investigation or the solution of the wrong operations research problem.

In order to actually apply systems analysis to the problems of developing fishery policy it is necessary to appreciate systems methodology. This methodology involves a collection of concepts which I will try to summarize. The first concept involves the description of a system. A system is a collection of interactive entities among which flow the information or material. Because the real world is extremely complex, the analyst must obtain simplified but not necessarily simple views (see Park, 1962:1370) of the flow of information or material among the components of the system. This necessity generates systems which are, of course, abstractions of the real world and thus are artifices of the imagination. The cleverness, success, or failure of an analyst can be measured by the degree to which his abstractions of real-world systems contribute toward the efficiency of the analysis.

In addition to being an artifice of the imagination each system has further properties that include its inputs, its outputs, its resources, its environment, and its management mechanisms (see Churchman, 1968). The resources of the system are those entities that are modifiable by the system, whereas the system environment (not to be confused with the term "environment" in ecology) consists of those entities that are, with respect to the system, fixed constraints; these fixed constraints are unmodifiable by the system. For example, one fishing system might consist of a fleet of 10 boats which are considered to be fixed in number and 30 fishermen who must be allocated 3 to a boat. In this instance both the boats and the fishermen constitute part of the environment. An alternative system might consist of 10 fixed boats, and 30 fishermen that can be allocated 0, . . . , 30 per boat. For this alternative system, the boats are still in the environment, but the fishermen have become part of the resources of the system. We can see that even with a fixed number of fishermen and equipment, we can arrive at, even in this simple example, many alternative systems. (Many analyses tend to be deficient because they form a rigid boundary between the resources and the environment of the system.) A final property that should be discussed is the way the system is managed, that is, how decisions are made in the system. These decisions can be made by people, as in the conventional view of management, or by machines or man-machine in the cybernetic view. Even if we are not engaged in systems analysis *per se*, should we not be able to identify these properties of the system that we are dealing with?

Thus given a problem situation we can see that its solution rests in attaining certain objectives which can be attained by the construction of alternative systems, each having inputs, outputs, resources, environment(s), and management. The analysis attempts to identify all possible alternative systems which might achieve the objectives and then choose that alternative system or set of alternative systems that will enable the transmission of appropriate advice to policy-makers. How should the best alternative system be chosen? It should be naturally chosen in a systematic way. A further fundamental concept of systems analysis involves this systematic way of choosing among alternatives. This is outlined by Quade (1968). I have modified his approach and have outlined it in Figure 5. It is not possible to go into details in this short paper. It is important, however, to distinguish a few particularly important aspects of the essence of systems analysis which appears in this figure. The first is that the analysis is iterative by nature, that is to say, if the interpretation is unsatisfactory, the analysis cycle is re-initiated at the formulation stage. Instead of time phasing a one year project to reach the interpretation stage at the end of a year, we would, using a system approach, perhaps schedule for the same problem several iterations during the year. We might want to sketch an entire problem from the formulation to the interpretation in a matter of months, thus making several passes during the year. It is clear then that in systems analysis we treat our objectives as a variable. This is only reasonable because after we make our first iteration, we know more about the problem than we did before. Therefore we should expect our view of the problem to change with additional information. Again, we are not only attempting to solve the problem as it is initially posed; we are also attempting to simultaneously determine if we are asking the right question.

A second feature of Figure 5 that bears emphasis is the segment on criteria. A criterion is a test that we apply to a set of alternatives to adjudge which alternative is, in some sense, best. It is important to distinguish criteria from objectives, which are missions that are to be accomplished. The substance of many analyses lies in the definition and reasonableness of criteria. I have discussed the nature of criteria and pointed out the need to develop economically related proximal criteria for fishery management (Rothschild, 1971).

The weakest link in many analyses is in the criteria which are used. There are a variety of common criterion errors. The commonly used cost-benefit ratio, for example, only makes sense in the unlikely situation (unless we are fortunate enough to be in a region of the problem

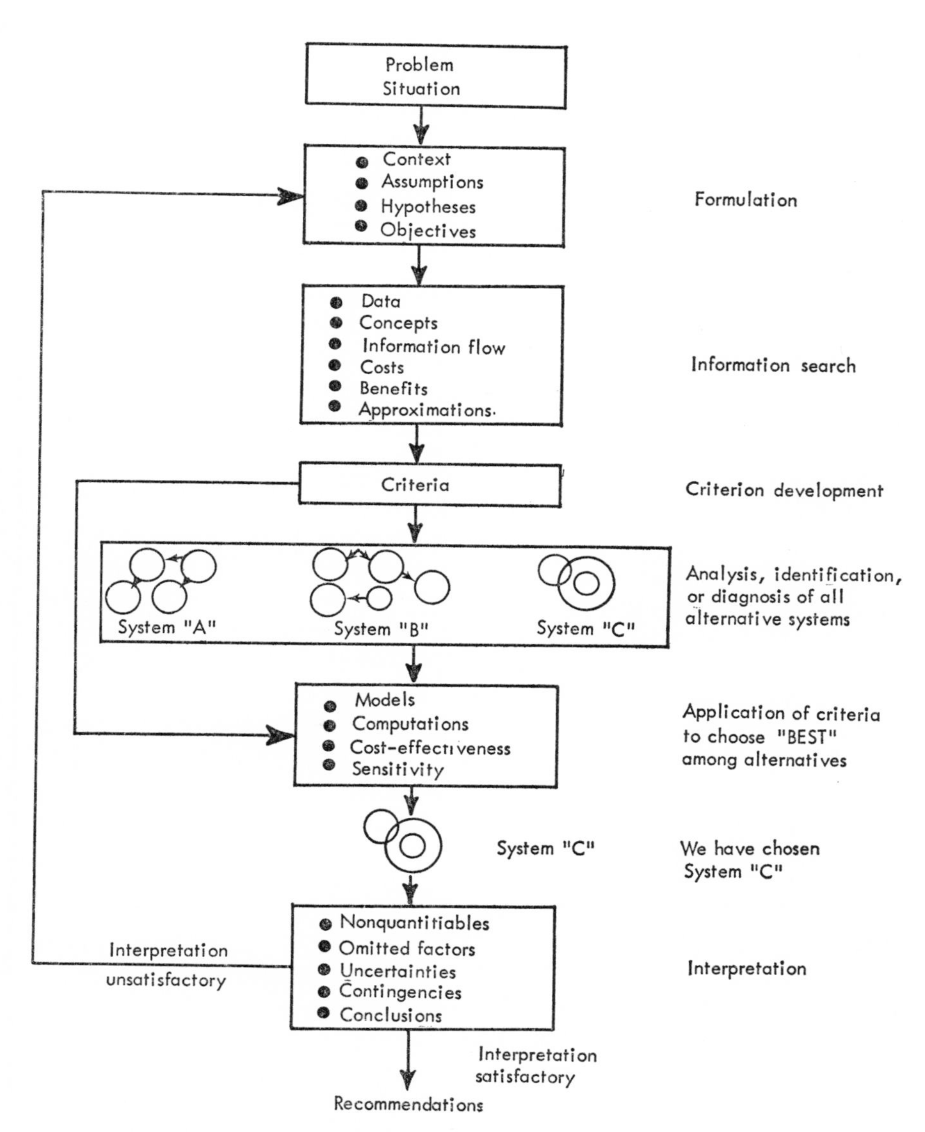

Figure 5. Steps in a systems analysis (after Quade 1968)

that is approximately linear) in which the costs and benefits are a straight line passing through the origin. Another misuse of ratios is the following (see McKean, 1964): We have two alternative research cruises, Cruise *A* and Cruise *B*. Cruise *A* will yield, all other things being equal, 1,000 temperate observations (benefits) for $2,000 (costs), but Cruise *B* will yield 2,000 observations for $3,000. The benefit-cost of *A* is ½ and of *B* is ⅔. Thus, in terms of benefit-cost, the system *B* is better than *A*. But is this an appropriate criterion? A better criterion for judging between *A* and *B* might be to ask the question of whether it makes sense to pay $1,000 for 1,000 additional observations.

Another criterion-related problem which may challenge the intuition is the casting of priorities (see Hitch and McKean, 1960). We are frequently asked, "What are the priorities?" We must rank activities in some sort of order and we attack the first ranked activity first. The vagueness of uncautious use of priorities can be seen in a list of desirable items; for example, a new car, a typewriter, and a pair of shoelaces. Many people would assign priorities to these items in the order that they are given. If, however, we mention in considering this "shopping list" that we have a budget of only $50, the priorities will instantly be reordered. Cost is frequently not included in setting priorities. A final example of the criterion problem is the maximum sustained yield criterion that has been and is commonly being used in fisheries, but is generally recognized to be a rather poor criterion. An example of the economic deficiencies in the MSY model have been rather clearly elucidated by Waggener (1969). An example is shown in Figure 6.

Thus the development of criteria is an extremely important part of systems analysis. While this activity should not dominate a study, it should be an important component, particularly in an area such as the applied aquatic sciences where the criterion problem has been given so little thought.

In addition to the concepts outlined in Figure 5 there are several others that need to be emphasized, even in a brief discussion. The first of these is the concept of suboptimization. We can view a system as a series of hierarchies. As an example, consider the food problem in country *X*. The analyst will try to contribute to the solution of a particular food problem in country *X* by breaking the problem into component parts. One component might be a fish sector and another might be an agriculture sector. Attempts to optimize either the fish or the agriculture sector are called suboptimization. Suboptimization can be dangerous if the suboptimalities are not consonant with optimality in the entire system. If we were to optimize the fish sector we could conceivably allocate considerable emphasis to meal production which might not be consonant with the solution of the food problem. On one hand, it is almost always

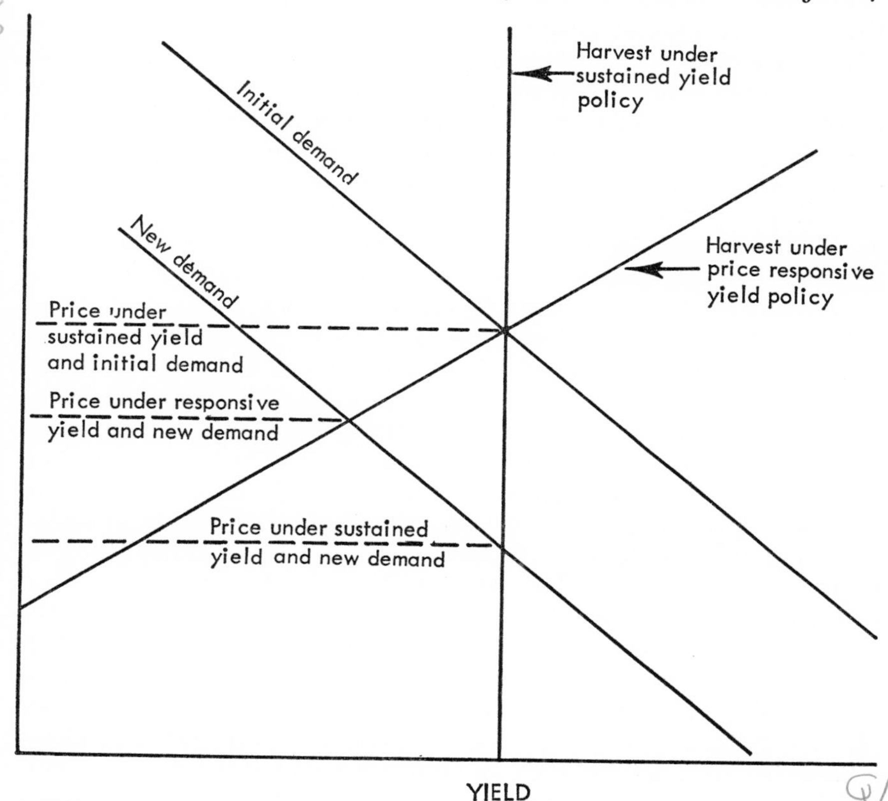

Figure 6. Possible relation between harvest and demand showing equilibrium prices under sustained and price responsive harvesting. The new demand creates new equilibrium prices and the fluctuation is less under a price responsive harvesting strategy than under a sustained yield strategy. (After Waggener, 1969)

necessary to suboptimize to make a problem manageable. On the other hand, the dangers of suboptimization must be avoided by making sure that the suboptimization provides optima that are constant with the main goals and objectives of the system. Many "solutions" to fishery problems are clearly suboptimal. The danger of suboptimization can be greatly reduced by taking, at least initially, as broad a view of the problem as is practicable.

Another concept involves the problem of overdetermined criteria. These are criteria that are set at levels that are too high to be of operational use. For example, a fishery research organization might want to promote the fullest use of stocks while minimizing all conflicts associated with the harvesting or processing of stocks. Is this even possible, and what information does a statement such as this convey to the employees and the constituents of the organization? This is clearly an example of overdeterminism; more proximal criteria need to be developed. A further example of overdetermined criteria may be various economic criteria such as attainment of economic efficiency. Advice on economic efficiency is frequently difficult to interpret operationally, because the advice, in itself, does not tell the fishermen, for example, how to be economically efficient. A set of operational criteria needs to be developed which are consonant with the higher level criteria of efficiency.

Finally, two components should be included in most of the alternative systems that could be conceived in passing through the process outlined in Figure 5. There is little use in a system analysis that does not include plans for communicating and implementing the analysis. In fact, many analyses are quite successful in one sense and failures in a more important sense. The success lies in their solution of the technical problems, but their failure lies in not planning implementation. A considerable portion of analysis, then, should be devoted to developing explicit plans for the implementation and communication of the system.

In addition to these concepts a variety of special techniques have become associated with the systems methodology over the years. While it is not necessary to be entirely familiar with these techniques it is useful to be cognizant of them when contemplating systems analysis. Knowledge of these techniques enables the analyst to appreciate the kinds of solutions that are possible for the questions posed. A certainly nonexhaustive list would include organization; technological forecasting; business decision theory; interpersonal communications; and computer methods.

With respect to optimization techniques, these can be found in any textbook on operations research. In addition to the well-known techniques of optimization of simple "nice" functions that everyone learns in calculus the reader will find various mathematical programing techniques for optimization of constrained objective functions. Most of these texts will also provide the bases for inventory and queuing theory. All of these have applications in fisheries ranging from fleet scheduling and management to research design.

The subject of technological forecasting is, as Bright (1970) says, an attempt to "assay the future." Systems analysis because of its philosophy places heavy emphasis upon understanding the future rather than describing the *status quo*. The subject of technological forecasting is difficult and fraught with incommensurables. Nevertheless, there is much about fishery technology that can be forecast. For example, if we take a ten-year planning horizon, what changes do we expect to see in jurisdiction, in vessel design, in fish detection, in fleet strategy, in refrigeration, and so on?

Business decision theory is taken here to include the techniques that are used to make business decisions. This includes such concepts as, for example, capital budgeting: the allocation of capital among investment projects where the capital could be fishing vessels and the projects could be fisheries and determining the present value of various investment schemes. The subject of program budgeting should also be mentioned to emphasize the need for fisheries work to concentrate upon programs rather than activities.

Interpersonal communication theory is, of course, a very large subject, but at least one aspect of this theory is particularly important to fisheries work and this deals with the use of experts, scenarios, and so on (see Quade, 1968). Who are the experts and what makes an individual an expert? How can we best utilize the knowledge and intuition of experts? One technique that appears pertinent is the delphi method, which was developed to obtain an opinion from a group of experts. The conventional approach is to have a meeting or a conference of experts and ask for an opinion or advice. The advice that will emanate from the meeting will most likely be colored by the most dominant personalities in the group—it may not be the best advice and it certainly does not represent an accurate view of what the group is thinking. The delphi method attempts to avoid this difficulty by developing a questionnaire on the pertinent subject. This questionnaire is sent to the experts and an individual is assigned to monitor and sort out the opinions; he then uses the responses to generate another questionnaire which is sent to the same group. The process continues until the opinions of the group are stabilized. The delphi method clearly involves a lot of work, especially since it forces the "meeting" sponsor to carefully identify and formulate the problem and objectives. The delphi method in many instances can be quite cost-effective.

Finally, it is necessary to mention for completeness the broad range of computer techniques which are frequently vital tools for the systems analyst. There are many texts on this subject and there is no need to dwell any further upon this particular aspect.

Thus, in conclusion, our thesis is that whatever the national policy has been with respect to fisheries, it has not been clearly stated. There is some question as to whether it is even possible to elucidate a policy owing to the diffusion of responsibility and concomitant diminished accountability characteristic of many more or less independent agencies involved in managing restricted aspects of our fishery stocks and one large agency that essentially has no administrative responsibility to manage the stocks. Even though this policy cannot be stated in terms of sets of decision rules with well-defined objectives, alternatives, and criteria, it is possible to see the effects of this situation. These include: (1) a more or less continuing erosion, or at least no increase in United States flag operations; (2) growth in the processing segments of United States fisheries.

Now, this may be a very desirable situation, or on the other hand, it may be undesirable. A penetrating analysis of this situation must be undertaken and I believe that the most efficient method for approaching this analysis is a systems approach.

In suggesting the need for a different type of analysis, there should be the opportunity for the diverse community of individuals—the fishermen, the processors, and other interested individuals—who interact with fisheries to guide the future utilization of these resources. I believe that the opportunity rests in having a stated policy from which the future consequences of any decisions can be determined.

March 1971

REFERENCES

Bright, J. R. 1970. "Evaluating Signals of Technological Change." *Harvard Business Review,* 48 (No. 1): 62–70.

Churchman, C. W. 1968. *The Systems Approach.* Dell Publishing Company. 243 pp.

Copi, I. M. 1953. *Introduction to Logic.* New York: Macmillan. 472 pp.

Hitch, C. J., and R. N. McKean. 1960. *Economics for Defense in the Nuclear Age.* Cambridge, Mass.: Harvard University Press.

McKean, R. N. 1964. "Criteria." In E. S. Quade (ed.), *Analysis for Military Decisions.* Rand Corporation R-387-PR. 382 pp.

Morse, P. M. 1970. "The History and Development of Operations Research." In G. J. Kellehez (ed.), *The Challenge to Systems Analysis: Public Policy and Social Change,* pp. 21–28. Publications in Operations Research, No. 20.

Park, T. 1962. "Beetles, Competition, and Populations." *Science,* 138:1369–75.

Quade, E. S. 1968. "When Quantitative Models Are Inadequate" and "Principles and Procedures of Systems Analysis." In E. S. Quade and W. I. Boucher (eds.), *Systems Analysis and Policy Planning.* New York: American Elsevier Publishing Company. 453 pp.

Rothschild, B. J. 1971. *A Systems View of Fishery Management with Some Notes on the Tuna Fisheries.* FAO Fisheries Technical Paper No. 106. 33 pp.

Waggener, T. R. 1969. *Some Economic Implications of Sustained Yield as a Forest Regulation Model.* University of Washington, College of Forest Resources, Institute of Forest Products, Contribution No. 6. 22 pp.

14

Science and Fisheries Management

DAYTON L. ALVERSON

Several years ago, in preparing a lecture to be given at the University of California at Los Angeles, I wrote: "Individuals, outside of professional conservationists, generally believe that the great majority of regulations promulgated to govern use of the living resources are based on conservation principles substantiated by technical information garnered on the status of stocks exploited."

State and federal agencies responsible for common property resources, however, often find it politically expedient to exercise their responsibility by managing the fishermen or the fishing systems rather than by managing the resources. Many of the fisheries regulations, both in the United States and elsewhere, are only remotely associated with conservation objectives and instead have their basis in resolving gear or economic conflicts—that is, problems of allocating resources between different user groups.

If one had to classify the basis for regulations governing the use of the oceans' living resources, both coastal and oceanic, status of stocks is the most frequent rationale given, although allocation of the resource and the economic status of users are frequently involved. The tendency to ignore or mask economic issues involved in management is rooted in the fact that the mandate of the management agencies is restricted to conservational aspects.

In the United States, the implementation of regulations may proceed as follows. A research arm of a state agency investigates an exploited resource and finds it to be overfished; that is, productivity of the stock has been reduced either because too much biological material has been

Dayton L. Alverson is director of the Northwest Fisheries Center, National Marine Fisheries Service, National Oceanic and Atmospheric Administration, United States Department of Commerce. He also serves as affiliate professor, University of Washington, and vice-chairman of the Food and Agriculture Organization Advisory Committee on Marine Resource Research. In the past he held positions as acting director, Bureau of Commercial Fisheries; associate director for research, Bureau of Commercial Fisheries; and director of the Exploratory Fishing and Gear Research Base, Seattle, Washington. Dr. Alverson has served as a delegate for the United States at numerous international meetings dealing with fisheries matters and recently was a member of the United States delegation to the Preparatory Conference on Law of the Sea.

removed in too short a time period, or because the parental stocks have been reduced to a point where recruitment has been impaired. The research arm of the regulatory agency may subsequently recommend regulation and, if adequate statistics are available, propose minimum sizes of fish that may be harvested coupled with some seasonal restriction on fishing effort or catch. If the director of the agency can convince his constituents that the management is in their best interest, a regulation concerning level of use may follow—even though many of the underlying economic facts that tend to reduce overall efficiency of operation continue to operate.

Quite often, however, regulations result from pressure brought about by one segment of the industry against another user group. Even where strong conservation measures are required, administrators frequently prefer to regulate the hardware of the user or to re-allocate the resource among the user groups in a manner that penalizes the more efficient units.

International management is plagued with similar problems. Although considerable lip-service is given to the concept of conservation and management of the resources, there is, indeed, a great deal more rhetoric than management. Although the key word in the international arena is again "conservation," the underlying problem often is the allocation of resources.

In examining the history of the use of the oceans' living resources, there seems to be several fundamental problems which have inhibited their effective management. They include: (1) awkward institutional arrangements for implementing management schemes, (2) failure of the scientists and administrators to differentiate the roles of science and management and to acknowledge the underlying economic factors, (3) failure of science to provide information which is timely in terms of management requirements, (4) adherence to a rigorous scientific method as a basis for making management decisions—in particular, a religious dedication to this concept in terms of management of international fisheries.

It is not my intent to deal in depth with these particular problems as they have been discussed frequently. It seems appropriate, however, to briefly discuss them.

In the United States implementing effective management of coastal fisheries has, as most people are aware, been impaired by the fundamental distribution of regulatory authority. For the most part, domestic fisheries are controlled and managed by state agencies. In theory, this management (or control) is restricted to those waters inside the three-mile territorial sea. However, the states have, for the most part, managed resources on the open seas by specifying the kinds, sizes, and amounts of fishes and harvesting gear that can legally be aboard fishing vessels when within waters under the states' jurisdiction. This has often led to a hodgepodge of different regulatory measures imposed on United States coastal fishermen, including direct attempts to exclude "outside" fishermen. Efforts to minimize differences through interstate compacts have met with varying degrees of success but have not solved the basic problem.

At an international level, authority for management either resides in commissions or is implemented directly by the federal government as a result of bilateral or multilateral arrangements made between countries. These institutional arrangements, however, frequently depend on committee action requiring unanimity of views by users. Perhaps an even greater problem is the failure of the institutions to define management goals and to secure the authority required to effectively manage the resources under their jurisdiction.

These problems are uniquely associated with fisheries in that fishing is the only remaining major food producing system in the world which relies on hunting or exploiting wild stocks, and must harvest stocks that are considered common property of all sovereign nations. These factors set fisheries apart from other major food producing systems and require the successful user of the oceans' living resources to deal with problems characterized by the (1) dynamic nature of living resources, and (2) common property aspects of the resources.

In a historical sense we have attempted to deal primarily with the first problem and for the most part ignored or dealt with the second problem in a de facto manner. The theoretical response of living resources to man's exploitation is well known, and management is generally rooted in the hypothesis that (1) one can maximize the yield from a certain cohort under proper utilization strategy; (2) the surplus yield is maximized at a population level that is approximately one half of that which prevails at the virgin stock condition; or (3) there is some predictable parent-progeny relationship which if understood can be used in conjunction with yield per recruit theory to establish an optimum harvest strategy. The last may be considered to be exclusive from a stock recruitment theory, although the mystical response of the Schaefer model (see, e.g., Schaefer, 1954) to exploitation changes requires an understanding soul.

A reluctance to understand or consider the allocation and economic problems in management decisions has resulted in considerable confusion. Indeed, as conservation is a universally accepted noble objective, it is usually given as the reason for certain unilateral actions or bilateral or multilateral arrangements between nations. The conservation rhetoric, however, often masks the fact that the arrangements are frequently based on economic rather than conservation issues.

To manage the oceans' living resources effectively, we will have to recognize the wide range of social-economic problems which are the product of their common property aspect. Although more and more resources are being overfished as a result of the common property problem, we do not seem to have either the vision or the legislative or institutional capacity to resolve the problems. Most professionals recognize that overcapitalization, in terms of capacity required to fully utilize living resources, results in extreme political pressure on management agencies to soften the nature of regulations or to adopt regulations which eliminate or restrict the more efficient harvesting units in favor of the less efficient units. The result of this type of management is to degrade the harvesting system and not to provide the physical protection the stocks require.

Another factor that has inhibited the adoption of timely management has been the failure of scientists and administrators to differentiate between the roles of science and management. Gulland (1971) notes that some problems in terms of efficient management seem to be "due to the confusion between the roles and methods of working of science and management." Management is a matter of making decisions and it is often as important to make a decision in time as to make precisely the best decision. Management has to resolve a wide range of political, social, or economic problems. Science has to provide evidence on the likely results, within its field of competence, of possible management actions and so enable more rational decisions to be made.

This, apparently, is not well understood; and, indeed, it would appear that many institutions and scientists have decided that decision making in fisheries management should be subject to the same demands of rigorous statistical testing as the general scientific mode.

In commenting further on the problem, Gulland states:

> It was a fallacy to think that scientists, given time, and perhaps also money, can produce the complete answer to management problems; e.g., specify the precise value of the maximum sustained yield from a particular stock of fish, and also the exact levels of fishing, and of population abundance required to produce it. The inability of scientists to produce such complete and exact analyses has resulted (sometimes deliberately) in a delay in the introduction of management measures.
>
> Scientific finality cannot be achieved. Science advances by disproof rather than proof—a succession of hypotheses are put forward capable of explaining the observed facts, and have to be abandoned or revised as further observations show them to be inadequate. Complete and final scientific advice cannot therefore be provided; all that can, or should, be provided, is advice that is sufficiently accurate and detailed for the immediate purpose.

It is further apparent that complete scientific understanding is not necessary for effective management. Perfection is almost impossible to achieve in fisheries matters. A single goal of attempting to achieve optimum physical yield, even if it were possible to attain, seems rather pointless. Rather the aim of the institution should be to consider a wide range of possible actions and to take steps which seem to offer the greatest long-term benefits to the local or global community.

I think it not an exaggeration to state that a fair part of the world community is not satisfied with existing arrangements to resolve fisheries problems. This was clearly reflected at the Preparatory Conference on Law of the Sea held in Geneva in March 1971. The difficulties in using existing legal arrangements provided by the 1958–60 Geneva convention for resolving fishery problems were pointed out by Ambassador Jorge Castenada of Mexico. The ambassador stated that the rights provided by the existing convention were almost useless. He continued:

Do you want proof that this is so? In the 13 years that have transpired since the Convention was consummated, this new and important right [coastal state preference and unilateral implementation of conservation] has not once been exercised in any part of the world. There exists, however, a great number of cases in which the coastal state has been obligated to tolerate excessive abuse by foreigners along its coast. . . .

The anxiety of many coastal states concerning the endless debate over status of fish stocks was well stated by Mr. J. A. Beesley, the Canadian ambassador, as well as many other delegates to the March conference. Mr. Beesley stated:

While we realize the complexity of the problem, we are nevertheless wary of some of the highly complex remedies that have been proposed in the past, and which may be proposed for consideration by the next Law of the Sea Conference. From the point of view of a coastal state, any proposed solution which entails endless discussions by fishery scientists who, however objective they may be, find much room for disagreement because fishery science has not yet become a precise science, is not a satisfactory solution to the immediate and urgent problems of a government in protecting the livelihood of its fishermen and the industries dependent on fishing. Even if the fishery scientists come to agree on scientific assessments, the administrators representing their governments in any commission or other regulatory body that may be set up may not accept the recommendations of the scientists, and the governments themselves may not accept the recommendations of their administrators because of political pressures. Any complex proposal based on proof by a coastal state of economic necessity for its industry, or on preferential rights based on amount of investment, on sharing of quotas, etc., will involve endless disputes which will be difficult to settle, while in the meantime the fishery resources of a coastal state will be disappearing.

These two quotes reflect dissatisfaction with existing procedures for managing and allocating living resources of the sea. The problems are obviously more easily identified than is establishing more effective institutional arrangements for their resolution.

In this sense, let us speculate on possible future arrangements for allocating between users the oceans' living resources. I think it fair to state that the status quo will not prevail regardless of the outcome of the 1973 conference; that is, the disenchantment with existing bilateral and multilateral arrangements and those afforded by the existing Geneva Convention are obvious. This discord, in addition to strong nationalistic tendencies, suggests that if the conference is to succeed, it must provide for more control by coastal states over living resources inhabiting their adjacent waters. Should this action be taken, the problem of managing the cosmopolitan, high seas species, such as tuna, marlin, swordfish, saury, and so on, would have to be resolved by some other type of control.

Let us speculate on possible modifications of international law that could affect management of the oceans' living resources, then ask ourselves whether or not the speculated changes would tend to resolve problems which have historically plagued effective management.

There are a myriad of potential schemes that might be proposed for managing the oceans' living resources, but most represent variations of several basic themes. Commonly discussed possibilities are (1) maintaining existing legal arrangements but strengthening coastal state preference—that is, preference for resources on which it has a dependence; (2) maintaining existing international legal regimes but establishing a system of international quotas for those species or species subgroups for which there is heavy international competition; (3) internationalizing the management activity for all marine species in waters seaward of the coastal states—some authority might be invested by the international organization in coastal state government; (4) extending the coastal state jurisdiction based on a zone extended seaward from its territorial waters; and (5) allocating responsibility for management to national, regional, or international bodies depending on the distributional features and behavior or the resources involved—that is, an ecological approach in which coastal states would manage their anadromous species and those which are generally confined to the neritic waters adjacent to the continent, and regional or international bodies would manage pelagic species.

However, let us suppose, perhaps idealistically, that the purpose of the Law of the Sea Conference as regards living resources is to establish a new legal order for managing resources and the behavior of man's activities on the oceans. This new order would facilitate more harmonious relationships between users of the oceans' resources and would establish institutional arrangements to allow for timely implementation of measures designed to maintain the productivity of the resources. If we examine the foregoing possible arrangements, it becomes apparent that the first two principally attempt to resolve allocation problems. The first would do so through greater recognition of the coastal states' economic interest in adjacent resources, and the second by establishing national quotas, presumably through some arbitrative procedures. The first arrangement recognizes the special interest of coastal states in resources off their coasts, and the second recognizes the increasing demand by the world community on limited resources. The latter would attempt to provide stability to existing fisheries and insure participation through guaranteed access to a portion of the resource.

Both the first and the second options would do little to alter existing institutional arrangements, nor would they necessarily improve factual information on which management decisions are based. The problem of how to handle developing nations which deserve to start their own fisheries—generally those with low catch quotas but large ambitions—would be difficult to resolve, and if not resolved could lead to continuous bickering and discord among all parties concerned. The quota system, unless evolved with other control mechanisms, could lead to economic waste in utilizing the oceans' living resources.

Internationalism of the oceans, from the standpoint of managing their living resources, would constitute a substantial change in the institutional arrangements for fisheries management. In theory, it might allow management schemes which encompass the spectrum of social-economic and biological factors influencing the productivity of exploited resources and the economic welfare of its users. Allocation of resources between user groups under international schemes might, however, be a perplexing problem with overriding political pressures influencing decisions of the super regulatory agency. An effective mechanism for acquisition and analysis of fisheries data and implementation of timely policies for resolving problems would be a prerequisite to the success of international schemes. It is, however, difficult to imagine that a super bureaucracy could operate satisfactorily unless considerable authority were delegated to regional and local fishery bodies. Such bodies would have to operate within an internationally accepted management framework.

Extending the jurisdiction of coastal states would obviously resolve allocation problems, but not necessarily to the satisfaction of all users—particularly nations with distant-water fishing fleets. It also could provide for implementation of timely management. On the other hand, one should recognize that the effectiveness of this arrangement would vary from state to state and the character of management would largely reflect national attitudes and policy. In some states internal politics and pressure groups could have a major influence on resource management, and hence coastal state regulation might be no better than that which exists under present institutional arrangements.

Adoption of a policy of management based on ecological characteristics of species has certain advantages in that it would place a total resource in the hands of one management group—that is, a coastal state, group of coastal states, and so on. The state or group of states would have a strong local interest in the welfare of the resource and would not spread management of single populations or stocks between different management entities.

The allocation problem could be resolved if a policy were adopted which gave priority to coastal states for utilizing resources in adjacent waters which they were capable of harvesting. This would have to be supplemented by an arrangement for managing high seas pelagic species through international or multinational control.

Problems alluded to earlier concerning difficulties in obtaining timely scientific information and a proper understanding of the roles of science and fisheries management will not necessarily be resolved through adoption of new institutional arrangements. They require a changed attitude on the part of the managers and an understanding on the part of the lay public that precise biological forecasting is extremely difficult and costly to attain. Further they require that the scientists and the managers themselves understand what each can contribute to effective management and that "management" requires a decision or set of decisions and subsequent action if the consequences of not acting will result in depletion of the resources or economic loss.

February 1971

REFERENCES

Gulland, J. A. 1971. "Science and Fishery Management." *Journal de Conseil,* Vol. 33, No. 3.

Schaefer, M. B. 1954. *Some Aspects of the Dynamics of Populations Important to the Management of the Commercial Marine Fisheries. Bulletin of the Inter-American Tropical Tuna Commission,* Vols. 1 and 2.

15

Fisheries and the Quantitative Revolution

G. J. PAULIK

The titles of three books written since 1961 by Jay Forrester of the Massachusetts Institute of Technology's Alfred P. Sloan School of Management bear witness to the ever-expanding ambitions of computer simulation modelers: in 1961, he wrote *Industrial Dynamics;* in 1969, *Urban Dynamics;* and in 1971, *World Dynamics.* Forrester's work has been widely reviewed, not only by major scientific journals but also by such popular magazines as *Time, Wall Street Journal, Fortune,* and *Playboy.* Increasing popular interest in the computer revolution and its possible effects on society is indicated by the appearance of lengthy articles dealing with these topics in such periodicals as *Life* and *National Geographic.* It is difficult to identify precisely how and where the science and business of fisheries fit into the entire spectrum of spectacular computer achievements. As one might suspect, practical applications of this new and highly lauded quantitative technology are greatly limited by inherent constraints and impediments in any long-established activity such as fisheries. Ultimately, the most exciting applications of computers to fisheries will lie in the planning and decision-making areas. In the immediate future, however, the most valuable applications will be concerned with information collection, analysis, storage, retrieval, and the use of operations research techniques to optimize harvester behavior.

Computerized management will require machines with highly developed and very flexible decision-making capabilities; they should be able to deal with uncertainty and to handle large, ill-defined problems. Marvin Minsky, director of a project on machine-assisted computation at

The late *G. J. Paulik* was professor of fisheries at the College of Fisheries at the University of Washington where he taught the graduate sequence in population dynamics. His primary interests were population dynamics, salmon research, and computer teaching games. He served as a biometrical consultant to all of the state fisheries agencies on the West Coast of the United States and also acted as a consultant to a number of international commissions.

the Massachusetts Institute of Technology, was quoted in *Life* as saying, "In three to eight years, we will have a machine with the general intelligence of an average human being . . . a machine that will be able to read Shakespeare, grease a car, play out his politics, tell a joke, have a fight. At that point, the machine will begin to educate itself with fantastic speed. In a few months, it will be at genius level and a few months after that, its powers will be incalculable." In support of his claims, Minsky can point to the existing electronic person called SHAKEY developed by the Stanford Research Laboratory, and the impressive performance of checker and chess playing computers that have adopted heuristic learning strategies to update and improve their playing tactics as they accumulate game experience. These machines are able to handle game situations that could not have been foreseen by their designers. The implications are obvious: we are approaching the development of computers that can write and modify their own programs and thus conceivably could evolve if exposed to environmental conditions that place survival value on appropriate traits.

Even if Minsky's claims are off by a decade or more, they are still incredible. We may be certain, however, of an exceedingly long lag between the development of this type of computer with its tremendous potential for management and its application to societal problems. In an area as diverse and fragmented as fisheries, the "people" problems associated with applying computers are tremendously magnified.

Everyone is familiar with the Peter Principle, which says that men eventually rise to their level of incompetence. There is a lesser-known companion principle called the Paul Principle, put forth by Paul Armour, Director of the Computer Center at Stanford University. The Paul Principle states that men become uneducated and obsolete in jobs they once performed quite adequately and that this happens most often when most of their duties are managerial and they are in an area undergoing high technological change. Such men will not feel comfortable in the presence of computers that are capable of playing office politics. Of course these principles affect all sectors of society, even the university, where the staffs and faculties have great reluctance to apply to their own problems the sophisticated and imaginative techniques they develop to help others. For example, at the University of Washington, we do not find the computer center applying computer simulation modeling and optimization to their own operations to make more effective use of the existing facilities, nor do we have an effective campus-wide time-sharing system, and so on.

Before the new computer technology can be employed, a welter of interface problems need solution. Computers will have to learn to communicate with human managers in languages so problem-oriented that even managers trained twenty years ago can understand what is going on. Although many knowledgeable observers agree fisheries are ideally suited for the systems approach, they also recognize the large human behavioral components associated with fisheries administration and management and the general rule that the larger the human component of any system, the more difficult is the task of developing successful strategies for optimizing performance. Yet, surely it will be much easier to identify and implement a policy to obtain maximum physical yield of shrimp from the Gulf of Mexico than to design and implement a policy to reduce the robbery rate in New Orleans. To predict by means of mathematical models the total economic impact of a change in tax policy is at least an order of magnitude tougher than to predict by means of mathematical models the changes that occur in the dissolved oxygen concentration in a stream following a change in the characteristics of the waste discharged into the stream.

Numerical modelers are learning that much of the difficulty in societal engineering lies in establishing something analogous to a laboratory to test theories in order to allow corrective feedback between the real and the abstract worlds. The sorts of high expectations that used to be reserved for the arrival of the cavalry are now directed at new teams composed of aerospace-trained systems analysts and their biological support staffs, who are becoming very interested in environmental problems in general and in fisheries problems in particular as their traditional sources of support dry up. The initial high hopes of their fisheries clientele will be substantially lowered when they begin comparing the precise and detailed numerical forecasts they demand of the behavior of a fishery system to actual system performance.

Despite these difficulties many of the industrial systems based on exploitation of a natural resource have attractive qualities for the modern system analyst. I would like to list and discuss some of these attractive properties of fishery systems.

1. They are complex and defy "back-of-envelope" type of solutions. Large computers and systems methods are required to mimic systems structure and dynamic interactions between components.

2. There is a great demand for techniques of aggregating system complexity into numerical quantities that are relevant to the decision-making process. Many resource management problems exist in an adversary setting where there is a background of responsible negotiation with some assurance that realistic model outputs will be given serious consideration in the decision process. For example, delicate negotiations among countries trying to conserve and rationally exploit stocks in international waters often revolve about fairly sophisticated mathematical models of the fish populations. Thus there exists a substantial demand

for computer assistance from the many fishery regulatory agencies now operating.

It is informative to contrast model building activities of the practitioners of the abstract science of ecology (as distinguished from the social engineering branch of ecology) to those of the fisheries scientists. In ecology, models are used to clarify and communicate ideas, while in fisheries science modeling is a pragmatic activity directed at obtaining operational, acceptable answers as cheaply and quickly as possible. Mainly, the types of questions posed by fisheries scientists are: How many fish and what kinds should be taken in this season from a particular fishing ground?

Considering the true mundanity of fisheries management, it is amazing that it should generate any mathematical models at all. Of course, many of the standard models used today are nothing but elaborate methods of fitting curves to data. Although it might seem unusual that anyone should take these models seriously, the regulatory agencies in many cases have all the choices of a drowning man. Man cannot control the marine ecosystems, and any really extensive aquacultural activities are in the far distant future. Comparatively speaking, terrestrial systems are simple to manage, since man can modify these systems to certain monocultural configurations to produce plants and animals of use. In the sea, man can only harvest in harmony with the laws of nature. If he does not maintain a certain ecological balance, he destroys the productive capacity of the resource. Thus, fisheries scientists have been forced into ecological modeling because of the demands placed on them by resource exploiters. The grim effects of poor management are all too obvious. The ghosts of several fisheries provide the necessary documentation and case studies to illustrate the effects of poor management. I do not think anyone argues with the necessity of managing with extreme care marine mammal populations such as the Antarctic whale or the Alaskan sea otter and fishes such as spiny dogfish with life histories similar to those of the mammals. However, it is only recently that it has become so abundantly clear that such fish as the clupeoids, which were thought to be insensitive to the effects of human harvesting, in fact require exceedingly careful management if they are to persist as viable resources. The Norwegian fishery managed to reduce the Atlanto-Scandian stocks of herring enough to drop the catch from 1.2 million metric tons in 1967 to 0.2 million metric tons in 1969 in spite of increased fishing effort. Between 1964 and 1968, the Norwegians spent over one hundred million dollars modernizing their existing fleet and building new purse-seiners.

3. The data base is overwhelming and fisheries organizations are now experiencing a dire necessity to consolidate data and to transfer them

into usable information, that is, into knowledge. Our modern technology is inundating us with irrelevant data of every sort. Environmental monitoring systems are becoming more popular with government agencies. These enormously efficient data collection schemes—including satellite censors, and robot monitoring networks of all sorts—are producing in a wide variety of environmental circumstances data streams that defy containment. In my opinion, the most effective filter yet devised is an integrated simulation model that can be employed continuously in an ongoing sensitivity analysis to separate the relevant from the irrelevant and to pinpoint needs for high resolution.

4. Serious study of resource behavior under alternative management schemes requires the time compression possible in computer simulation models.

5. Resource managers need to identify statistically vulnerable hypotheses before running expensive field programs, and use of appropriate computer simulation models may be the most feasible means of making such identifications.

6. Many fishery resource systems are far too valuable not to model before running even the simplest type of field experiment. Of course this is the same reason that we play World War III games on computers. Thinking about the unthinkable is one way of preventing it.

7. There is a great need for a universal communication medium in international negotiations, and mathematical models to some extent express problems in a universal language.

8. There are now many superproblems in fisheries management that require super-problem-solving capability, for example, pulse fishing by distant water fleets has wiped out the lead time formerly available to develop management strategies as the fishery evolves.

As one might expect, the potential of computerized simulation modeling has been recognized by many other groups. In fact, at times I get the feeling that a significant portion of the biologists in this country are beginning to believe that mathematics is the *only* language for expressing ideas about biological processes and that we are currently witnessing a complete turn-around from the traditional descriptive qualitative approach in ecology. We must now guard against overquantification. The possibility of premature and forced quantification of all sorts of biological phenomena is evident in the spending policies of the major agencies and foundations, for example, the International Biological Program has turned into a large-scale experiment in ecosystem modeling. The National Science Foundation's Analysis of Ecosystems Integrated Research Program is now funding six major biome studies. Two of these, the Grasslands Biome and Deciduous Forest Biome, are well under way, and four others—the Coniferous Forest Biome, Tropical Forest Biome, Tundra Biome, and Desert Biome—are in the process of starting up. NSF is also funding a major modeling study on biological productivity of marine upwellings. The National Oceanic and Atmospheric Administration is currently discussing funding a series of major estuarian biome-type projects involving considerable numerical modeling. IRRPOS (Interdisciplinary Research Relevant to Problems of Our Society), which has now been absorbed by the RANN (Research Applied to National Needs) Program, is currently supporting several interdisciplinary groups such as the one headed by Ken Watt at the Davis campus of the University of California. Watt's group is trying to understand societal needs for energy in the state of California by means of model building. Successful execution of his project requires modeling, as starters, land use and the growth of the human population in California.

Public reaction to the environmental crisis has stimulated modeling and also research activities in a much broader field related to environmental management. Constraints to modeling are not widely recognized in the current fun-and-games stage of development. Karl Banse and I listed many of these constraints in "Biological Oceanography: Models," a 1969 article we wrote for *Science* (13:1362–63). These constraints are particularly applicable to the field of biological oceanography:

1. There is insufficient quantitative knowledge about functional relations such as the manner in which demographic characteristics of a given species change with the population density, or how the efficiency of food assimilation depends on environmental factors.

2. Useful data for building detailed, comprehensive models, and for validating them, are scarce. The obvious way of directing the proper collection of data is through stimulation and encouragement of model building, which will point to the data most needed.

3. There is no agreement upon resolution required for a realistic ecosystem model. At one extreme are models of gross energy or mass transfer between trophic levels and at the other, models that rival complexity of nature itself by including numbers of individual animals of each species.

4. Biological oceanographers with only a few notable exceptions have made little advance in synthesis of available knowledge. Marine ecosystems are so complex and available information is incomplete, that many people may regard prospects from meaningful syntheses to be poor. There is also lack of examples demonstrating significant achievements with complex ecological models in biological oceanography. . . .

5. Development and improvement of simulation models are usually multidisciplinary efforts, but there seem to be few people who can serve as leaders for such efforts.

6. It is doubtful that experience in ecological modeling is sufficient to support development of an ecological computer language, although a general

language may be useful in the future. However, computer subroutines to assist biologists in modeling some general ecological processes would probably increase the use of simulation modeling.

The first constraint is self-evident; many more observational and experimental studies are needed at combinations of various levels of interacting factors to define essential biological processes adequately. The second constraint is also concerned with assembling the types of essential data needed to fuel a realistic simulation model of a fishery resource. The third constraint is an inherent difficulty in any model dealing with animals with vastly different generation times and life histories. One possible solution is to employ a continuous system world view but to evaluate numerically the differential equations used to represent the system at times determined by phenomenological criteria. Models with interconnected sectors at many different levels of resolution could be focused as a microscope to study the detailed behavior of part of the system imbedded in a rich background of related activities modeled at low resolution.

Figure 7 illustrates the cones-of-resolution approach to modeling an

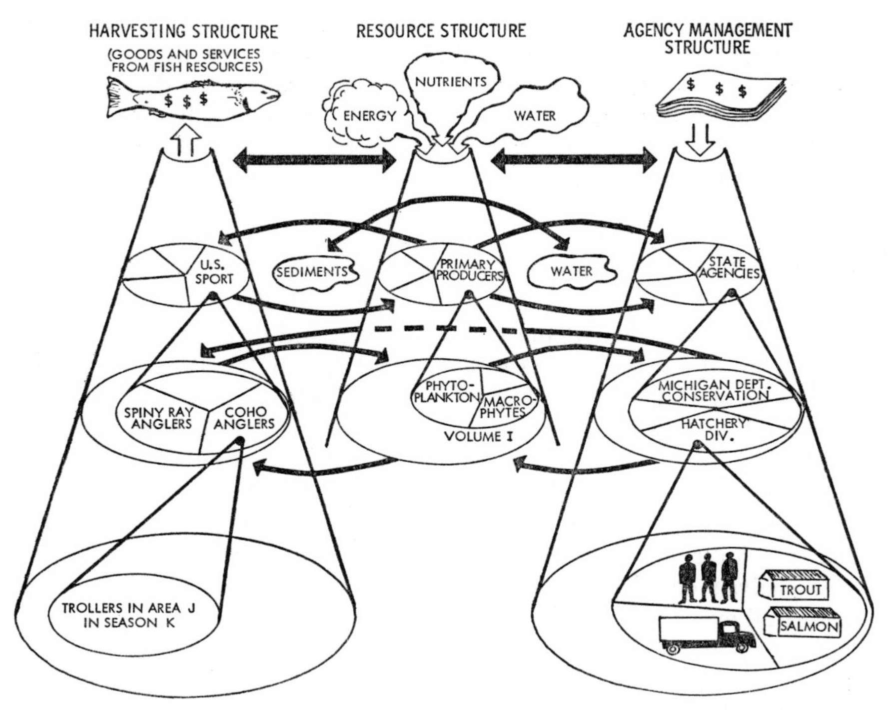

Figure 7. Cones of resolution (the higher the level the coarser the representation)

extremely large fisheries system. Each level in the cones of resolution shown represents a degree of complexity. At the apex, the grossest level, cash flows into a management agency structure to interact with available natural resources and a harvesting structure to produce goods and services in the form of fish. Resolution is increased by working down the cones. Individual groups of animals, individual governmental agencies, and individual harvesting gears become visible as we descend. The details of small operational units are available near the bottom. Note that models can work at a high level of resolution on one cone and simultaneously at low levels on other cones.

Constraint four will disappear if any of the biome exercises, for example, are successful.

Constraint five is an important constraint. Integration and communication requirements of the large interdisciplinary teams needed to construct realistic ecosystem models were vastly underestimated when the NSF began its Analysis of Ecosystem Modeling Studies. Much difficulty can be avoided by a business type of management structure; however, in many cases such a structure is incompatible with the tradition of individual excellence in the biological sciences. Models need exceedingly long lead time as well as continuous evaluation during all phases of construction and subsequent use. There are very few dedicated individuals who have the stamina to fight through all of the difficulties required to see a large-scale model develop to the stage where it is useful as a basic resource of planning. The simple technical problems associated with debugging are far greater than most people realize.

There is substantial disagreement about constraint six. Many computer specialists feel that FORTRAN is still adequate for the task and that there is little to be gained from developing general ecological languages. However, it is difficult not to be impressed with the acceptance of statistical techniques that followed development and distribution of general package routines such as the Biomedical Computer Programs of the University of California at Los Angeles. My guess is that development of general ecological languages especially oriented to critical processes which determine population productivity would move the field of fisheries management forward by making new ecological procedures available to resource managers at all levels of competence. Computer centers could be linked so that highly specialized libraries of routines for fisheries analysis could be stored at one center and be accessible to other centers all over the world.

Large-scale simulation models should accept data in as natural a form as possible. Management agencies may be nature's bookkeepers but they are staffed with men who chose to become biologists and management agents, not accountants. A model or a pre-model data conditioner must accept historical data in the form in which it is found in the investigators' file cabinets. Our primary understanding of fisheries is in terms of hypothetical "statistical" fish; models must allow management personnel to think about their problems in terms of the entities with which they are familiar. Modelers who accept data only in the format of such academic concepts as energy flow, diversity indices, and stability will not be welcomed by management agencies. This is not to deny that we do need to develop a variety of relevant metrics. I certainly expect to see many ecological metrics evaluated and tested in field situations; eventually, if found to be useful, these will be adopted into everyday usage by the management agencies.

A critical problem that we have not yet mentioned is that of interfacing—how do we interpret several waist-high stacks of computer output sheets? Graphic and summarization techniques and new methods of presenting the results of computer analyses are required. It is a simple matter for a computer to jam all of a human interpreter's input channels; it is far more difficult for the computer to display and communicate its results in an understandable fashion to a human decision-maker. If computer specialists are going to influence political decision-makers, they must be concerned with precisely this problem. One possible solution could be extensive use of the new portable terminals with time-sharing access to some distant large-scale computer. Some lightweight terminals are equipped with display screens and will respond to a light pen and can be operated from your favorite congressman's or senator's desk. This type of terminal allows the congressman to use a computer in his home office to answer some familiar questions.

Data system capability is a people-related problem; every agency wants its own unique data system. Interchangeability is usually given second-order priority. I would hate to see our data managers running our railroads.

It may be asked if we should even bother with fisheries as part of the computer revolution. After all, fish constitute only a small part of the total flow of energy in an aquatic ecosystem. There are several reasons, however, for an affirmative answer to this question. First, fisheries has a successful modeling history. Many of the international agencies use models extensively to formulate management policies, and by and large modeling agencies have been successful agencies. Second, fish have high political leverage, for example, it is doubtful United States laws controlling pesticide use would exist at all had not the coho become heavily poisoned with DDT in Lake Michigan where they are most visible to the nation. Third, fish are important in absolute terms for their chemical properties, e.g., a recent NAS/NAE report on waste treatment shows man adds 1.8×10^5 tons of phosphorous to the ocean annually. Fishermen extract approximately 60 m.m.t. of fish from the sea which contain 20,000 tons of phosphorous. Fisheries transfer significant amounts of protein from one part of the world to another. Peru, for example, takes 10 m.m.t. of anchovy and produces 2 m.m.t. of fish meal to be shipped to Northern Hemisphere countries for feeding swine, poultry and cattle.

Before closing, I would like to mention the impact of computers on an area of fisheries which is most exciting to an educator. By letting students play computerized resource management games in which they participate by assuming various roles, they can acquire in these "link trainers" management experience that would take hundreds of years to accumulate in a responsible management position in the real world.

In conclusion I would like to reiterate two points: (1) the computerization of perhaps most human managerial activities during the seventies and eighties is inevitable; (2) both the business and the science of fisheries are participating actively in the computer revolution—fisheries has long enjoyed a reputation as one of the most quantitative of the biological sciences. I predict our reputation as quantitative biologists will remain untarnished and may even be enhanced during the next two decades.

January 1971

16

Management of the Exploitation of Fishery Resources

G. L. KESTEVEN

To be able to make a contribution to a seminar and book dedicated to Wilbert M. Chapman and Milner B. Schaefer is a privilege especially to be esteemed by any fishery scientist. In paying my respects to those two, who played such distinguished roles in recent fisheries science, I must express a hope that what I have to say will be found worthy of association with those bright names. But if there was a characteristic (apart from their common interest in fisheries) which Chapman and Schaefer shared it was, for my money, their utter fearlessness in taking a position and making known the view they held from that position. I am sure they would have urged me to take courage and present my views, even while they honed their own intellectual knives for an exercise in cutting-down-to-size.

Very briefly, what I have to say here is this: events in the fishery world over the past few decades have shown us that our natural sciences account of resources does not give us an adequate understanding of the behavior of those resources, so that, in consequence, most changes take us by surprise and only rarely are we confident that we are acting for the best—any best—in our utilization of resources. Yet even more forcefully those events have shown us that our strategy of resource use is not a matter merely of how a resource reacts to what we do. There are technological, human, and social elements to be taken into account, and for these we need a social sciences contribution. This analysis, obvious as it must be to many, has led me to suggest that we must recognize a distinction between a technical plan according to

G. L. Kesteven is the manager of the United Nations Development Program/Special Fund Fisheries Research and Development Project in Mexico. He has served as fisheries advisor to the Government of Mexico; assistant chief, Commonwealth Scientific and Industrial Research Organization, Division of Fisheries and Oceanography, Australia; regional fisheries officer in Southeast Asia, and later chief, Biology Branch, Food and Agriculture Organization; and fisheries advisor to the United Nations Relief and Rehabilitation Administration.

which exploitative activities would be guided, and the political arrangement for allocation of opportunity to participate in those activities. But, in order to put that distinction into effect—to bring precision to our guiding of exploitation and thence to carry conviction in our arguments for the political arrangement—we must expect, I argue, an overthrowing of our present resources research paradigm and some dramatic change in our attitudes to resource utilization. I have set out in what follows some ideas on what these two revolutions may mean to us.

The New Responsibilities of Management

The exploitation of fishery resources is, and always has been, managed, but this is not to say that it is well managed. We tend to suppose that to manage affairs is to have effective control of them, yet when we manage to get into trouble we do not think that we are in control: on the contrary, trouble happens because we have lost control. Nevertheless, when we speak of management of exploitation we do generally intend "effective control of exploitation" even if we are not entirely clear in our minds as to the nature of the control, its effectiveness, the objectives toward which it should function, or by whom it should be exercised.

Our purpose, political in the fullest and wisest sense of that word, is a fishing regime, for each resource, which will serve community objectives as to the product to be got from the resource and as to the costs of equipment, manpower, and so on, to be spent in getting the product. Restraints are set upon community objectives by the characteristics of the resource, but even more precisely, restraints are imposed on the realization of the objectives by the behavior of the operatives of the industry, and, above all of course, by the behavior of the fishermen.

In discussing the kind of regime we might envisage for a particular resource I wish to suggest that it can be established and made effective only if each participating fisherman, fleet, and nation not only observes the rules of the regime, but participates in the formulation of its objectives and contributes to operation of the mechanism for designing its rules, modifying them as may become necessary, and enforcing them. That is, the fisherman must not be able to continue to regard regulations as some alien restraint imposed upon him for purposes he does not recognize, for reasons he does not understand, by persons whom he does not know, for interests which to him appear to conflict with his own. He must be brought to such level of participation that he would no more contemplate anti-regulation action than a farmer would think of pulling his crop of fruit green and utterly unmarketable. It is our responsibility to persuade him by the evident rightness of our arguments, to convince him by the plain evidence of results, and yet to put him in the way of deciding upon matters for himself and of contributing to community decision because he can understand how the situation suits him. We cannot eliminate the fundamental risks and uncertainties of work at sea, nor require of a fisherman that he act other than he may see fit to do with respect to those dangers. But we can provide him with services to help him avoid some proportion of those incidents; we can improve the equipment with which he faces those dangers; and we can assure him of the ready availability of rescue when necessary.

Our more difficult task, however, is to provide an effective apparatus of management which, employed for the totality of community objectives, nevertheless will assure each fisherman of some measure of security in this most improbable of worlds, while offering him some possibility of the unexpected in this so highly regulated world. From that apparatus the fishermen, the industry, and the community should be able to expect (1) guidance in development, as to kind and quantity of equipment in all sectors, and as to the relations between elements, and (2) guidance in the conduct of operations. With the management of exploitation we are concerned with an operational regime, bearing upon resources, within an overall system of management of affairs. Here I want to refer to another division of our subject matter, this time at the level of community affairs: the distinction between what, in shorthand, I shall call technical plans (for example, a fishing regime) and those matters which, in similar shorthand, I shall call political arrangements, such as allocation of opportunity to participate in a fishery. I propose in the course of this paper to discuss the nature and importance of this distinction.

The Structure of a Fishery

For the purposes of this discussion I shall refer to Figures 8–10 in which Robert Ingpen and I have tried to represent the totality of a fishery system. The system comprises two major parts: the biotic elements (systems) which are the resource, and the technico-socio-economic elements which are what we most commonly think of as the industry.

The biotic elements (Fig. 8) operate as they do of their nature; in exceptional cases, employing our practices of biological research, we may learn something of that nature before exploitation starts. In this state our efforts have negligible effect on the characteristics of the system and we may learn enough to be able to describe the behavior of the system, in its relation with its contextual system. Our description may be so successful that from observations of particular elements of our biotic system and/or of its contextual system at particular times we shall be able to predict, qualitatively and/or quantitatively, the charac-

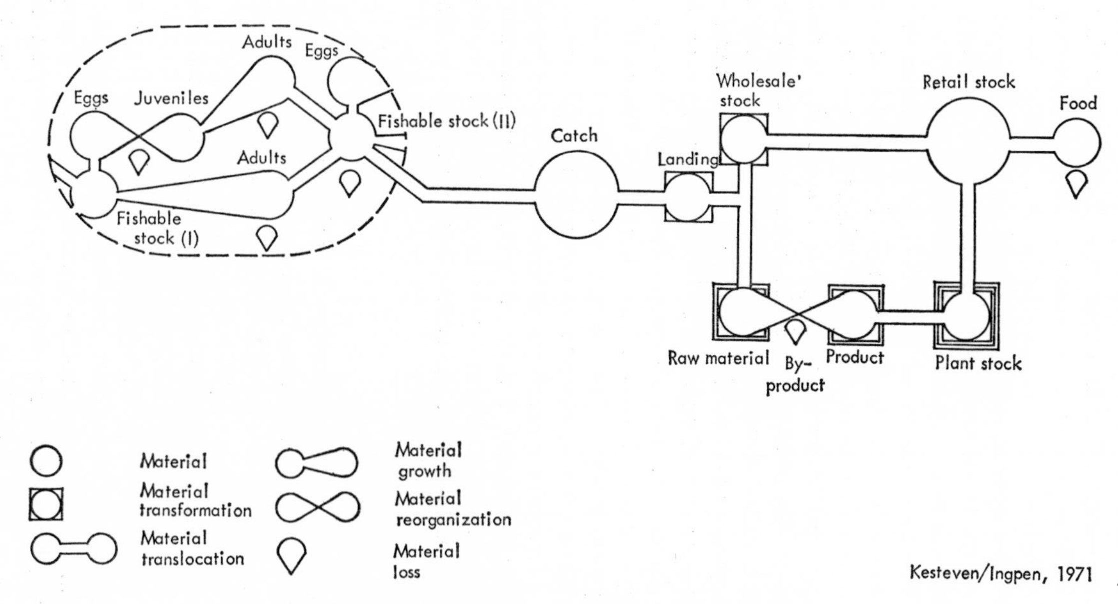

Figure 8. Fisheries structure: materials flow

teristics which our biotic system will have at some other time, in the future. As soon as fishing starts (Fig. 9) our efforts become a significant factor in the operation of the biotic system and contribute to determination of some of its characteristics. Our intervention can be at various levels, and of various kinds, and it is with various levels of confidence that we can predict its results.

The socio-economic elements (Fig. 10) are thoroughly contingent in the sense that their structure and operation are determined by human decision. I propose not to consider the possibility that we have no free will, and that all our decisions are determined.* I therefore assume that we are concerned with decision processes which determine the structure of this technico-socio-economic system, and that when we talk about management of exploitation we are talking about those decisions. What I propose to do then is to examine the nature of the decisions we must make with respect to action over various parts of this system, and to examine those decisions with respect to the information on which they are based, the principles according to which they are formulated, and their consequences.

Resource Use

The entire orientation of our thinking is to something we call "resource use," an expression which obviously has meant several different things. It can mean merely that which we do with what we take from a resource, as swallow an oyster which we have prised out of its shell, leaving the lower valve on the rock to which it is fastened; but it can also mean all those other, progressively more complicated processes by which we convert an aquatic organism into a form suitable for ingestion, including feeding it first to other animals (which is the basis of Chapman's *bon mot:* "If you ate eggs for breakfast this morning, you have already eaten seafood"). In this latter sense the principal classes of fishery resources use are:

1. of its products
   - A. directly, for human consumption
   - B. as food, for cattle and poultry and as fertilizers for food crops and hence, indirectly for human consumption

* We do not yet have the means (if we ever shall) of knowing the deterministic sequence, supposing it to exist, and while we do not we must act as though it did not exist. If, however, we did know it we would be able to decide to act differently from it; if a very determined philosopher were then to claim that that decision too would have been determined, we could discover the sequence that led to it too and then depart from it, and so on in an infinite regress, looking for the original design. Thus, either a deterministic sequence exists and can be known, and being known can be caused not to exist; or it exists and cannot be known and then, as Wittgenstein says, "What we do not know we must pass over in silence," and it is as though it did not exist; or truly it does not exist.

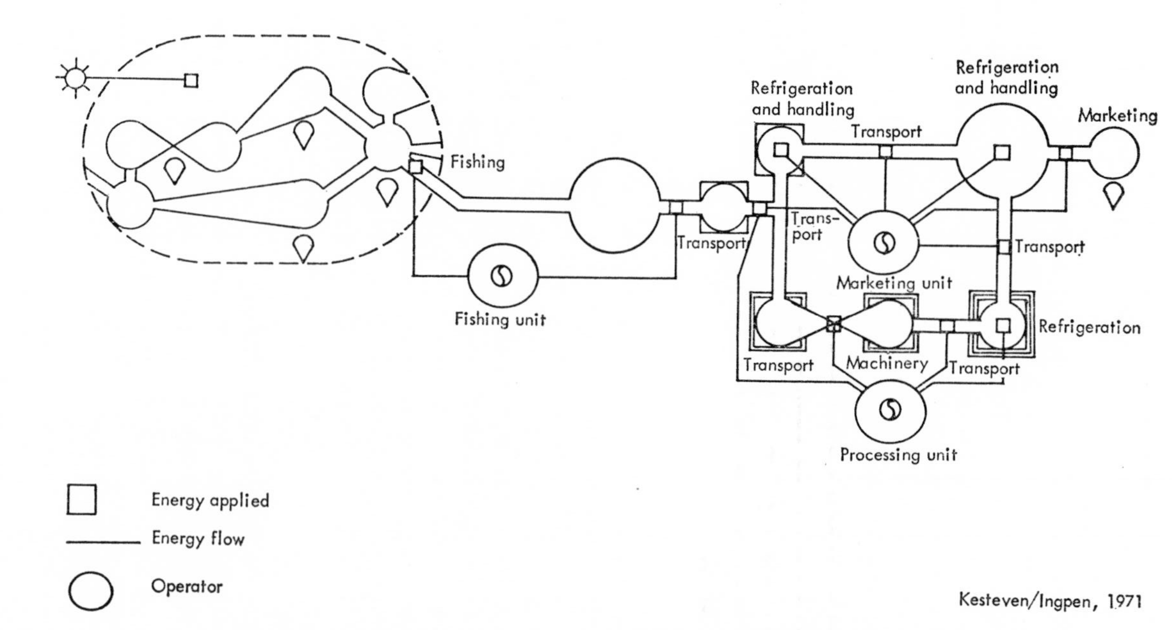

Figure 9. Fisheries structure: energy flow (including materials flow)

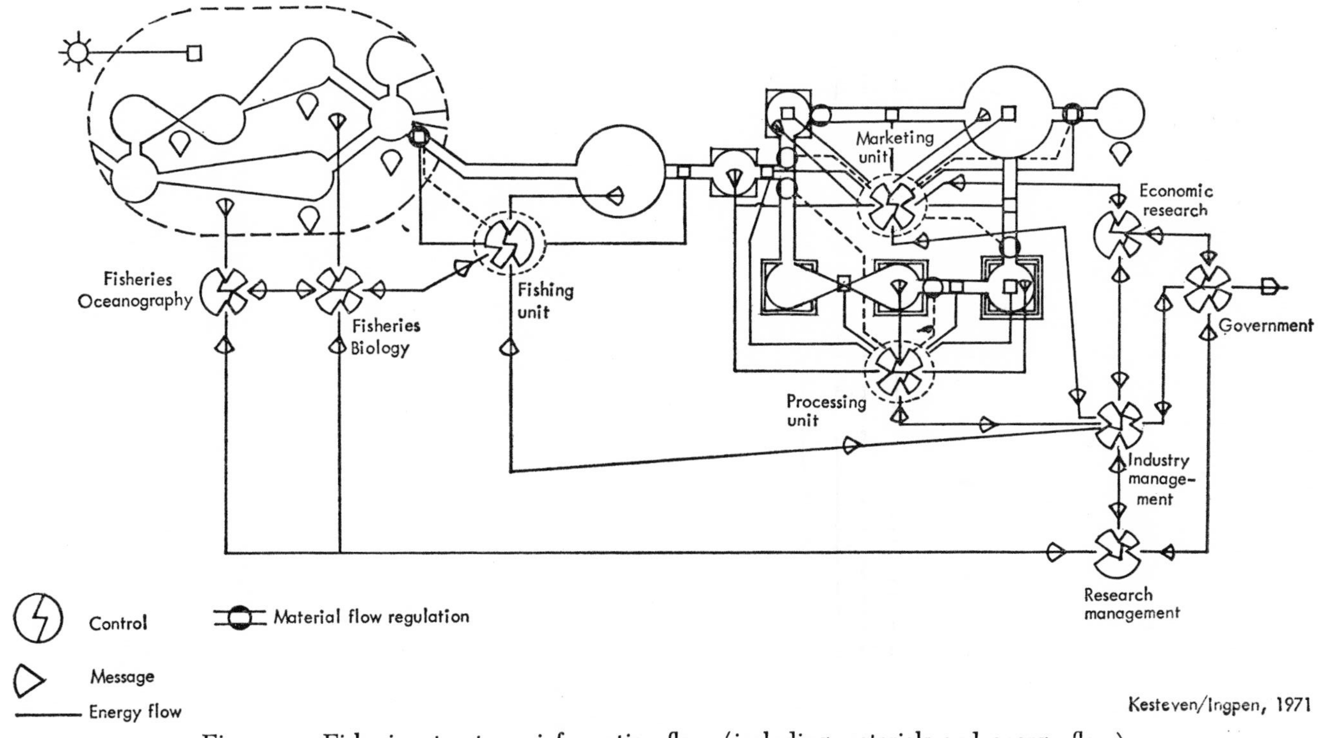

Figure 10. Fisheries structure: information flow (including materials and energy flow)

2. as object of recreational activity

From these uses flow numerous consequences and benefits, in employment, trade, and other economic activities, and it is of these as well as of the quantities of different products that we must take note. Of the quantities of product we must also consider dietary significance.

"Resource use" can also mean the whole scheme of technology, taking account of and evaluating all the practices by which we obtain materials from a resource and convert them. Further, the expression can mean economic intention, represented by various indices of absolute and relative levels of use. For any resource-use plan, whether relating to a single stock or to all of these resources of the world, we construct indices of use: these may relate the amount of use (product) of one year to the cost of securing it; and we may construct a third-level index relating the value of one of its relative indices in one period to its value in another period. Finally, each meaning we give to this term assumes a special significance for each of the systems to which we may have it in mind to apply the word "resource."

Remarks on the nature of resources must normally be supererogatory in this place, but some are necessary in order to identify and show clearly the premises of some of my later propositions. I do not mean the simple connotation, "what can be used." What I am after is the set of distinctions between (in our case): (1) resource as fishable stock, (2) resource as population of, say, juveniles and adults of a species in one place, or of all the individuals of a species constituting a unit stock, (3) resource as population in either of the preceding senses plus the food supply to that population, and (4) resource in ecosystemic sense, comprising population (specified as to its niche) and its habitat.

These are four entities, or better let us say "systems," which differ in extension and in physical characteristics, and which therefore are to be described by different sets of information, to be represented by different models, and their behavior to be reported, analyzed, predicted, and controlled by different practices. Any biologist of reasonable perception might at any time during the past hundred years have specified these four categories of resource, and have forecast some of the differences of technology and exploitation that their own structural and functional characteristics implied. Yet, the refinement of the fundamental concepts of these systems, and the forging of models to represent them, has been a protracted process, and the conflicts between fishermen (rising to importance at the international level) and the difficulties encountered by administrators in their search for effective means of resolving those conflicts both suggest why the process has been long and point to the importance of pursuing it to its end. These conflicts and difficulties classically show how much the management of practical affairs relies on the reach of scientific understanding of a managed system, a condition often denied by men of practical mould. I propose to examine this proposition by a brief historical review.

Resource Use Models: Technical Plans

When resource means fishable stock our exploitation is a mining operation. Description of resource thus defined was the work of, among others, Einar Lea and W. C. Hodgson on herring, Michael Graham on cod, C. F. Hickling on hake, and especially of Harold Thompson on haddock, all of whom studied life history and particularly composition of catch; even now the *Anales Biologiques* presents, as the result of research, tables of size and age composition of catch. This work is the base of the surplus production model, of which the data are: catch per unit effort (c/f) as evidence of density, catch (c) itself and effort (f); but c/f is not simply the ratio of c and f. While a system continues to behave as it did during a period in which some data were taken (no matter how this behavior resulted from interaction of the system with its setting) predictions made from the model are likely to be fulfilled, but if the system changes substantially, the predictions are likely to be wide of the reality. The model may be useful in stable situations in which perturbations—"momentary" and restoratory, or tendentious—are rare, but it is unlikely to be useful in an unstable situation and in one driven to change.

Therefore, in my view, the surplus production model is relevant only briefly, and in only a limited sense, to any management situation and is at risk of being misleading. It may serve very well as a means of arriving at a first estimate of upper limit of catch, but unfortunately, considering the nature of the data for this model, the course of events may have gone so far, and investments may be pressing so hard, at the time that prognostications are made from it, that these may be no longer sufficient. At present the so-called quota of the Inter-American Tropical Tuna Commission is set at 140,000 short tons. Over the past three years the total has gone up from 100,000 to this figure, and in earlier years the model authorized the commission's staff to nominate even lower figures as the amounts which could be taken and which ought not to be exceeded if the recuperative capacity of the stock were not to be endangered. Yet 141,000 tons were taken in 1970 and at the beginning of 1971 the commission was able to record that it appeared safe to take 140,000 tons and even perhaps to go beyond that to 160,000 tons. Now, over the past ten years in which restraints have been imposed with various degrees of effectiveness, either the stock has had the capacity to produce up to 140,000 tons or perhaps more, or there has been a progressive change in the stock over these years. The model could neither detect nor predict any change, if change there has been. But if we are only now reaching the current potential the simple result is that the model has cost the nations some hundreds of thousands of tons of tuna, worth some millions of dollars. I am not attacking the commission and even more I am not attacking its scientific staff: I am only supposing that a relatively small proportion of the value of the uncaught fish would probably have enabled the staff to develop a more effective model. Moreover I would emphasize that this model offers us only a specification of total catch as a guide to management.

The development of concepts of resource as population and of methodology for its study in this sense is the achievement of William Ricker, of Raymond Beverton and Sidney Holt, and of many other workers. These workers gave concrete expression to a distinction between fishable stock and recruit stock, developed the methodology of instantaneous rates of M and F, and applied Ludwig von Bertalanffy's equation for growth. Harold Thompson had laid part of the foundations to this work, with his efforts to correlate brood strength with catch, but he did not look in the opposite direction, to correlate brood strength with size of reproductive stock. Ricker has contributed much on this matter. The B and H model essentially offers method to make best use of growth potential—not to predict recruitment or protect reproduction and recruitment.

It seems to me that the situation is as follows. A eumetric fishing diagram tells us, for particular value of t_0, K, $L\infty$, and M, the size at first capture from which we shall be able to exploit the stock of highest catch-per-recruit rate; but this does not tell us what the amount of the catch will be. Next, we can tell from this information and from the exploitation index (F over F plus M) whether we can take more or less than we are taking at present recruitment rates; but the predictions we may make from this apparatus are valid from the range of recruitment situations represented in the recorded data, and a new situation outside the range of our record is likely to find us unprepared. I understand that something of this sort must have happened a few years ago in the North Sea haddock stocks. Therefore, or so it appears to me, the approach through which we shall make our next important advance is the study of the stock/recruit relationship. This is not a novel view: it has been voiced frequently in recent years and was represented by the special International Council for the Exploration of the Sea meeting on the subject which took place in 1968.

However, it seems to me less important at this time to know whether there is a maximum to the stock/recruit curve than to determine the range of lower level of stock abundance where a possibility lies that an unfavorable set of habitat characteristics will initiate a serious tendentious change. Perhaps I am forced to this view by what seems to me an urgency in the exploitation of shrimp resources. As far as I know, the records of all shrimp fisheries show very great fluctuations in catch. It can be thought that these fluctuations, even if chiefly of natural origin, might be reinforced in their downward movement by the effects of fishing, or, conversely, that their amplitude might be less if we could organize and manage the exploitation of each stock separately. There is reason to believe that the stocks themselves are separate, rising and falling in abundance independently of one another, and that as in the case of Antarctic whales, an overall target catch can lead to excessive concentration on particular stocks.

To take resource as population plus food supply is to move deeper into the total system and to challenge yet another of the assumptions of the deterministic model, namely that the terms K and $L\infty$ of the von Bertalanffy equation have fixed values; and this entrains with it still further challenge of assumptions as to constancy of M and as to the operational unimportance of variations in K.

Ivlev (1961) has discussed the methodology of studying nutrition in its ecotic mode, and workers of various countries are studying the consequences to the methodology of population models of a variable instead of a fixed input with respect to growth, recruitment, and mortality characteristics.

The management implications of these possibilities are considerable. We may imagine that in the future our work will have given us precise identification of unit stocks and such knowledge of each stock as to permit us to specify a flexible regime for each, and to develop systems of monitoring and prediction to which the industry would be expected to respond. In seeking to employ such a procedure of management we would be obliged to relate the costs of obtaining its necessary information to the benefits that might be obtained, giving careful attention, of course, to the question of whether the operations of the vessels, singly and as fleet, could be managed with flexibility enough to take full advantage of such a system.

Obviously resource as ecosystem is the deepest and most extensive concept of all, and equally obviously it implies more intensive and more extensive research, more precise description, more reliable prediction, and, in the long run, more deliberate and disciplined action by industry in its exploitation of wild stocks. It implies also direct intervention to determine structure and process, as to magnitude and rate.

An intervention situation already exists in some degree, where fish culture is practiced, but this is scarcely what we have in mind. Science-fiction writers are already writing of, and the comic strips are depicting, a biotechnology far beyond our present capacities, but even these do not achieve the full comprehension of what will be. Undoubtedly there will always remain some stocks in which we shall make no direct intervention and for which our strategy will be of guided rational exploitation at the second or third level of resource definition. There will be others which we shall herd so as to reduce mortality, to promote growth, and to preserve accessibility and vulnerability. Here it is important to emphasize that such a biotechnology will not have as its objective merely to increase the rate of exploitation; instead it will aim at changes in the stocks, to our advantage, and reduction in the cost per unit production. Then there will be other stocks in which, in addition to versions of the foregoing practices, we shall intervene to influence reproduction and recruitment. We can easily imagine suitable practices to be employed in coastal lagoons and estuaries and even in relation with some resources of the littoral and continental shelf whereby we shall shelter and feed the juveniles or even maintain hatcheries so as to supplement natural recruitment with the objective of maximizing the utilization of the resources upon which an exploited species depends. Finally, there will be stocks for which we shall have total cultivation practices and while I do not need to elaborate the characteristics which these practices have today, I must say I believe that in the future the bio-technology of fish culture will be considerably further developed than the form we now have.

Resource Use Management: Political Arrangements

There are two features of this history I have so sketchily reviewed, to which I wish to draw particular attention: first, the debility of our regulatory systems; second, the prevalence in our work of the propensity for panchresta. What in fact do we have by way of regulatory measures? Specification of total catch, of composition of catch, of times and places of fishing, and of some features of fishing gears. But, how useful are these, as guidance to fishermen, or as means to achieving our objectives? And how do they relate to the array of decisions which a fisherman must make? A total catch by itself merely invites unlimited competition. The other measures either seek to reinforce a total catch limitation and intensify competition, or aim at making the best of the species growth potential and/or at protecting reproductive potential. As to the second feature: a panchreston is a word we utter to cover our ignorance. When we observe some phenomenon, knowing it to be the result of several undescribed processes, we label the result with our panchrestic word, as though it referred to one or other of the underlying processes. The term "fishing power" and an expression such as "generate fishing mortality" have something of this character. Modern technology is enabling us to analyze the fishing operation into its component processes and to measure each separately, and before long we shall be able to define fishing power as to what it is and not as to its effects, two or three steps removed.

What I am stressing in this review is a parallel between, on the one hand, the progressive refinement of our knowledge of resources and development of our procedures for prediction (and here I mean prediction of the behavior of natural systems and prediction of the result that we can get by intervention) and, on the other hand, the progressive development of discipline among operatives and of skill among managers. This parallel is represented by the evolution for each resource of a most effective regime of operation. This regime comprises the total catch or yield which may be drawn from each resource, with specification of the composition of that catch and then specification of the equipment and methods to be employed in taking the catch and the patterns of operation as to times and places of taking the catch. This regime must be specified as precisely as possible, tailored for each resource, and flexible so far as can (may) be operationally practicable to effect changes to meet changes in the natural system which we cannot control. This regime is required, in the first place, so as to achieve greatest efficiency both in utilizing the resource and in managing the inputs of effort and investment. It is required so as to reduce the risk to the fisherman and to serve as a basis for at least reduction of the conflicts which arise between fishermen. The last is now becoming of critical importance at international levels. It is not long since it was possible for the captains of a fleet to feel confident that when the resource they were exploiting no longer gave profitable returns they could move off to resources elsewhere. This, which was the colonialist attitude of previous centuries, is scarcely possible any longer. I believe that we already have more fishing equipment in the world than we need for the catches that can be taken from the resources in our present modes of exploitation. This proposition is certainly true for a large number of identified resources. It was excess equipment that created the situation that now prevails in the exploitation of Antarctic whales and it is excessive equipment which is causing the grave situation in exploitation of the yellowfin tuna in the eastern and tropical Pacific. It might once have been possible for a fleet to say: "It does not matter if our equipment is excessive for exploitation of such and such a resource and has the result that the fishing season is very short each year, because we can move off to operate on some other resource." But now when a fleet moves from one area in conformity with this plan of work and sets course for another area it passes on the way a fleet leaving that other area with the same proposition. The future then requires not only specification of fishing regime for each resource but agreement on the pattern of regimes for all high seas resources; with that it logically follows that there should be agreement on the complement of equipment to match that pattern of regimes. This proposition can be made even only on the grounds that construction of excessive equipment is itself a waste of resources.

I therefore see the future for marine fisheries as follows: Technically, we shall have a pattern of regimes agreed upon in terms of the distribution of fishery resources and formulated from objective data with regard to the characteristics of the resource on one side and of consuming patterns on the other, with managed technological efficiency in between. After that there will be a system of allocation of shares in the catch and in the exploitative operations. This political arrangement will, in the first place, concede to the coastal state all the prerogatives of its territory over those resources which lie in its own waters; but in each case the exercise of those prerogatives will be related carefully to the corresponding regime for the resources immediately offshore. Management of this part of the arrangement will imply a distinction between those coastal water resources whose life cycle includes no phase beyond the limits of the coastal waters, and those which spend most of their life in high seas waters and pass only briefly through the coastal waters. With due allowance for these prerogatives a formula must then be found for allocation of the high seas resources. It seems to me obvious that such a scheme of allocation must be related to a principle of need and therefore would have to begin with a scale of population numbers in each country; these numbers would then have to be weighted by a series of indices of need or deficiency, for example, of nutritional deficiency, of unemployment, and of some measure of economic development.

The Necessary Revolution I: In Biology

I hope that what I said above will have made it plain that I consider that we have made about as much as we ever shall of the methodology of the fisheries biology that grew from, as I see it, Johann Hjort to Ricker and Beverton and Holt. Perhaps there is something of exaggeration in saying that the situation calls for a revolution, but I shall explain what I mean by that in the following discussion. First I would like to emphasize what seem to be unsatisfactory aspects of our present situation.

Fisheries biology has had many successes in many parts of the world. Its practitioners have discovered resources; they have described distribution patterns and delineated migrations routes, and have related these to habitat characteristics so that fishermen have been able to use this information in their fish-searching; they have made estimates of yield which have served as basis for development plane; and they have measured mortalities and growth so as to counsel administrators as to appropriate fishing regimes in the sense of approximate total catch, times and places of fishing, and minimum size at first capture (with related advice on mesh size and other gear characteristics). But in spite of these successes, fisheries biology, so it seems to me, is still only limping along, and is sadly inadequate to the needs of today. Many of its statements about potential yields are no more than the simple arithmetic by analogy that any intelligent observer might make; the fishing regimes it proposes are imprecise and feeble, and mostly backward looking—the IATTC (probably the most "controlled" of all international fishery situations) has for several years been looking ruefully back on progressive increases it had not foreseen, and perhaps for the next few years it will look back on a progressive decline it could only fear, not predict from any basis of fact.

In its inability to predict, fisheries biology sits squarely and sadly beside ecology. An ecologist can tell an administrator that if, into a river system, are poured great quantities of materials of a kind totally unlike those to which the native biota is "adapted," the river system will change; he can say little more. So it is with most other habitat changes man is bringing about: change there will be, must be, but unfortunately we cannot guide it, choosing each direction and justifying our choice by explaining our rejection of the reasoned alternative. This, it seems to me, is because ecology has run into a dead end of the kind of which Roger Revelle spoke informally during the University of California's "Perspectives in Marine Biology" meeting in La Jolla in 1956; he said that it seemed to him that most marine biology at that time was little more than laboratory physiology and histology practiced on marine organisms.

In the terminology of Kuhn (1962) (I hope I do not misrepresent him) fisheries biology is now operating with a paradigm whose potential of creativity is almost exhausted; employing its practices and models can serve only to confirm and refine the results already established; its practitioners are mostly engaged on convergent research. And this is where I find justification in talking of revolution: fisheries biology (and ecology—perhaps even biology itself) now needs divergent research; it needs someone to come up (as Ricker and Beverton and Holt did in the 1940s and 1950s) with a new set of practices and models and to challenge the retention of the old paradigm. It is my purpose here to make a small contribution to that revolution by indicating some aspects of the current paradigm whose revision might point the way to divergent research. I know that I shall draw the wrath of those who do not care to be cautious with words, and express a contempt for semantics. But this is precisely my particular point: I am advocating a revolution in our attitude to symbols, words, and semantics.

It does not seem to me an exaggeration to say that my case can be argued from the very evident fact that ecology is a field *par excellence* in which the principles of general systems theory can be applied with profit. The systems studied by ecologists are patently complex and manifest phenomena which are only slightly more comprehensible to the human mind than those studied by sociologists. It is a paradox that, confronted daily, even hourly, with ecotic phenomena we are less confident in our predictions of them than we are of much else of which we have less frequent contact. Or, to paraphrase a current tag, that "we know more about the surface of the moon than we do about the oceans," we may say that "we can predict the behavior of atomic systems with greater confidence than we can predict the behavior of any natural ecosystem." Although paradoxical, this situation is unsurprising, considering the variety of systems which had first to be inventoried by ecologists (this has been their main occupation for some time) and considering the nature of the concepts after which they are reaching.

The current state of the ecologists' vocabulary plainly shows the state of their work. A very great proportion of their vocabulary consists of apographic terms—names of things and situations; only relatively few terms deal with measurable processes. The many definitions offered for each of what appear to be important terms, and the ambiguity of many statements employing these terms, show that ecologists have not yet been able to make themselves sure of what it is they are talking about. Birch's (1957) remark, *à propos* of a claim that competition is centrally important in the dynamics of natural populations and in evolution, that "we should know what it is that is claimed to be so important," could be made with respect to many similar claims by ecologists. The present situation of ecology is that its progress is delayed pending the grasping of essential concepts, and in my view those concepts will continue to evade us until we purify the language of ecology. I suggest that some steps can be taken in this direction through the application of the principles of general systems theory and with special attention to the matter of symbols.

Ecology is the study of ecosystems; that is, it is that part of man's scientific activity that is concerned with organisms (singly or in numbers) in their habitat. The purpose of this essay will be served by examining some of the intellectual implications of defining ecology in this way: I suggest that these implications lie in many areas (or levels) of thought with respect to, among others: (1) the opportunities that ecosystems give for developing general systems theory, and the contribution the latter can make to ecology; (2) the theoretical constructions (models) by which ecologists represent these systems; (3) the operational strategies and tactics by which ecologists obtain confirmation of the validity of their constructs; and (4) the social responsibilities of those who engage in ecology.

Science, in its aspect of human activity, is one of man's responses to his condition. A plant is capable of only physiological responses, but man, with his mind (of which the evidence is convincing) and whatever else there is signified by the terms "conscience" and "spirit," is capable of retrospective and prospective action in addition to immediate response. He is aware of himself, as he was in the past, as he is in the instant, and as he may be in the future. Moreover, he is able, uniquely, to employ symbols. Without accepting any commitment to discuss the question of whether these characteristics were given man for the purposes of some superhuman being, we can accept the fact of their existence (however well or poorly we may apprehend their nature) and recognize that men can and do act in the light of this fact; that is, men do act purposefully in response to the circumstances of the present, in the light of experience of the past, and in some expectation of the nature of the future. In these terms I seek to declare my alignment with Bentley Glass in his writings (1956) on the ethical content of science, so as to argue that we may (and should) examine a sector of scientific activity not only with respect to the skill with which its practitioners elucidate the systems they study, but also with respect to their ability to put men in a position to do those things they need to do in regard to the systems being studied by that sector of science.

I believe these matters to be especially relevant to an assessment of ecology today because, with the vast increase in his population and the immense increase in his technological capacity, man is today making changes, in his habitat, of a dimension and kind never before experienced. The problems of food supply for human population, of pollution, erosion, and other deleterious effects on the habitat, and of urban spread, are matters to which, in principle and in practice, ecologists should be able to give real answers. Yet they can do little more than provide *ad hoc* schemes for the amelioration of the effects of human activity. Very little reliable prediction can be made from ecological results. Therefore, if it is intellectually interesting to examine the status of ecology and to scrutinize its bases, it is socially an urgent matter to try to bring ecology into some greater effectiveness through that examination and scrutiny.

Man's responses to his condition are artistic, religious, scientific and pragmatic. The first is concerned, perhaps, with "what might be" in a search for inner and enduring meaning; the second I suppose to be concerned with "what ought to be" if man were not mortal; the last is man's transaction of his daily affairs, working with and in the accumulation of the past, in some measure informed by what he has learned from his other responses. Science is in essence an intellectual activity projecting human understanding, by speculation, beyond the point it stands upon at the moment of speculation. Its demonstration is in logic and its confirmation in practice. In principle the speculation is in the form, "If I do so and so I shall observe such and such to happen." It is immaterial whether "so and so" is an experimental manipulation of a controlled situation, or a placement of an observer at some particular point in space and time. It is also immaterial that the proposition may be of the form, "If I were able to do so and so." The only requirement is that others should be able to repeat the performance (of logical argument in examination of evidence) and satisfy themselves within the canons of science that the proposition is sufficient ground for either pragmatic action, or further scientific speculation. It is this verifiability, made logically possible by the formulation of propositions in refutable terms, that characterizes scientific activity and distinguishes the scientific from man's other responses to his condition.

This verifiability, however, rests upon two principal bases: the validity (in some fundamental sense) of man's logical processes, and the validity (in the sense of "true correspondence") of the symbolic apparatus with which man effects his speculation. There seems to be not one single logic but several, and yet I would suppose that the invention of symbolic and later forms of logic may not be inventions at all, but only an opening up of more of the general logical matrix. Bronowski (1971) writes of a logic of the mind which is different from the overt logic we practice with symbols and pencil and paper, but I see no reason for supposing that what the mind does unbid differs other than in speed and quality from what it does under observation. Thus, I consider intuition to be logical process unobserved and I would grant its results the special respect only of its achievement.

As to the validity of the symbolic apparatus, we are here concerned with at least two lines of thought: first, the acuity of perception of the scientific observer; second, the skill, imagination, and inventiveness with which he finds symbols to stand for that which he has perceived.

While the former derives to some considerable extent from the training the observer has had, from the skill he has developed in practice, and from the reliability and dimensions of his memory, it also derives from that something labeled "knack" (of which Joseph Needham has written at length in his monumental *Science and Civilization in China*) which, like the exquisite balance of a ballerina, seems to be a superb coordination of the faculties developed out of natural gifts (above all, of the power to concentrate), practice, and interest. Of these things each of us has some intimation from personal experiences and from observation of others.

When we turn, however, to the second, we find ourselves in obscurer realms where even von Bertalanffy, like Homer, can nod. I find in one of his essays the following statement: "A coin is a symbol for a certain amount of work done or food or other utilities available . . . ," but I find this quite unacceptable and to be a misconception of the nature of symbol of scientific purpose. A coin is in no sense a symbol for any of the goods that might be bought with it for there is no fixity of relation between coin and commodity; display a coin to person *A* at one time and it may, according to the circumstances, lead him to think of some commodity, or it may not; on a second occasion this person may give entirely different response to even precisely the same coin; or show it to person *B* and a still different response is possible. What is enduring (at least for a time) and of which a coin is symbol is a set of conventional arrangements and agreements with respect to some concept of *value* and to the transferability of goods and service in transactions intermediated by coinage. The word "token" probably conveys better the informational significance of a coin. By "informational significance" I refer to two kinds of information:

1. The conventional information that passes upon apprehension or awareness of a particular symbol. This information is potentially present in a symbol wherever it lies (of a coin; in hand, pocket, or vault) and is ready to be communicated to and received by anyone to whom the symbol is known. This information is in no way reduced or modified by any act of apprehension.

2. The particular "recoverata" evoked out of the apprehender's mind upon being confronted with the symbol, in consequence of: (A) the circumstances under which his mind became trained to recognize the connotation of the symbol; (B) his own state at the moment of apprehension; (C) the circumstances of the act of apprehension; and (D) accidental and adventitious features of the symbol. Thus the prolonged recall by Proust in response to the madeleine was entirely personal and unrelated to any of the properties of madeleine. None of the characteristics of a coin (such as its metal, by whom it was designed, its age, or the mint that made it) lies within the scope of that which it symbolizes, no matter how vividly these might come to the mind of a numismatist. As symbol it stands for what has been agreed upon, and this characteristic it shares with all symbols of any usefulness. Thus for a system of symbols to be of any value it is necessary to have agreement upon (1) what is symbolized; (2) the symbol to stand for what is symbolized; and (3) the rules according to which the symbol can be used to stand for what it symbolizes. A symbol such as a mandala, which is intended, having no

fixed referent, to unlock any number of doors of memory, and to release any number of emotions, has no place in scientific activity. By the same token (if I may be forgiven the pun) any word of itself, or any usage of a word, that can call from him who receives it more than one response is of doubtful value in science. William Empson writes of the "Seven Types of Ambiguity" in poetry, validly, with persuasion and conviction because the opening of memory's doors and the evocation of emotion are the functions of poetry; it may even be argued that in terms of the real experience of humans such multivalued responses have a deeper validity, but as far as we know, science cannot yet proceed with such symbols.

When, then, we come to examine any particular sector of science we have to examine the integrity of its employment of logic and the nature of its symbolic apparatus.

In general, scientific activity is one or other of two kinds: the apographic or inventory making, and the syntactic or "relation identifying." Beyond this is the philosophic activity of metascience (science of science; scionomics) that takes scientific activity itself as a field of study.

Apography is taxonomy, descriptive anatomy, descriptive ecology, chemistry, and, indeed, every sector of science insofar as it catalogues its systems. Syntactics on the other hand is each of these sectors insofar as it looks for relations (1) between the systems of which those systems it studies are composed, (2) between its systems themselves, and (3) between its systems in the relationships that compose the systems of the level above that of the systems it studies. Thus, histology studies (1) the relations between the cells composing the tissues that are the object of histology; (2) the relations between tissues; and (3) the relations between tissues in the relationship in which tissues stand to one another to constitute organs.

To demonstrate the significance I see in the difference between these types of scientific activity I must discuss the difference between, and at the same time the identity of, object and event, and similarly the two principal connotations of the word "process."

From among the various meanings attributed to the word "process" in dictionaries, two have currency in contemporary science. Process I denotes a sequence of "things" (cf. procession) while Process II denotes a succession of actions. The latter is well represented in the expression "a metabolic process," the former in the expression "the process of evolution." The difference between the two is plain: Process I, as represented in phylogenetic, ontogenetic, or ecogenic (ecotic succession) sequences, is, as it were, a set of "stills" from a cinematographic record; what is represented on each "still" differs significantly from what is represented on all others; indeed, each frame of the cinematographic sequence itself might show significantly different form, but, whether we take the representational set of "stills" (as being easier to memorize and write about) or every single frame (no matter how many frames per second) we are looking at *something becoming something else.* In contrast, Process II (which necessarily contributes to Process I as result) is *something causing something else;* an enzyme working on its substrate, as one of the processes by which food is broken down in digestion; the muscular contractions, shell movements, and other actions, by a scallop, whose consequence appears as ejection of water; the operation and muscular system by which a predator seizes its prey. In every case the agent and the recipient of action are separately identified and both are separate from the result of the action.

To confuse the two connotations of the word "process" is to run risks of many kinds, of which the most likely is that one will suppose that in describing a sequence which is Process I one is identifying its underlying causes, that is, its component Processes II. Neither the structural bases nor the dynamics of Processes II can be subsumed from observation of a Process I: they have in each case to be identified and measured. A prediction from observed Process I alone has no more force than "if the circumstances and conditions in which this process occurred persist without change (which includes no modification of the Processes II of which this process is expression) the process will recur and, considering the temporal relations of the several parts of the process, the reappearance of part *X* may be expected at such and such time." It is obvious that much of the regularity of our lives derives from the essential regularity of the systems with which we deal, permitting us to ignore the underlying mechanisms. But it is by observation of Process II that man assumes authority, and it is only by study, analysis, and measurement of Processes II of many instances of any particular Process I that we can arrive at an estimate of the reliability with which we can predict Process II from Process I. Thus, a description of some phylogenetic sequence tells us nothing about what caused the changes from one stage to another, and it is only begging the question to appeal to "evolution" as the cause, for this is no more than saying that "such and such a change took place because there were events of which the change is evidence." In plain terms, evolution as process never caused anything; that of which we speak when we use this word is only result whose significance is of our attribution.

Of course, Process I is a prior phenomenon in human experience: we first observe things to happen and later we look into that complex of which what we have observed is merely the superficial evidence. In this sense, the distinction between Process I and Process II is no more than the difference between stages of acquiring knowledge of any system. But it is, at the same time, the difference between apographic and syntactic scientific activity. For in apography we take note of results, of objects *per se,* whereas in syntactics we examine events. In this sense it was natural that taxonomy and descriptive science preceded analytical science.

However, my thesis, which concerns the state of ecology, is somewhat more than the simple proposition that ecology must become analytical, for it can well be asserted that a great deal of analytical ecology goes on today. What I am arguing is that ecology is not yet effectively analytical and that it is in this condition because of conceptual and linguistic problems which have their origin chiefly in a confusion between Process I and Process II. I further argue that in order to achieve the linguistic purification required in ecology (and to proceed hence to a methodological revision) ecologists must identify the basic relations with which their study is concerned and must acquire a suitable apparatus of symbols with which to represent (1) those relations, (2) the events that occur when forces act through those relations, and (3) the connection between these events and the consequential sequence (Process I).

An attempt at a scheme for representing biotic systems and the relations prevailing within them (but not the events occurring) has been offered by Kesteven and Ingpen (1966). Figure 11 developed from that paper proposes a set of symbols for fundamental active relations of biotic systems. These relations are to be distinguished from passive relations (of time and space) and from the conventional, legal, and other relations of human society. Ingpen and I contend that all that happens in biotic systems (below the level of human social organization) consists of forces operating through these relations. The operation of any force in any particular instance is of course distinguished as to (1) the nature of the entities at the ends of the relation; (2) the timing, speed, and intensity of the force; and (3) the consequence, at each end of the relation, of the operation of the force. Furthermore a relation does not hold in isolation. Always there must be a set of relations, the entities between which they stand constituting in such relationship a biotic system, and the operation of forces through sets of relations constituting the events that, viewed intimately, are Processes II or, seen only as a result, are Process I. Ingpen and I believe that through this or some similar scheme it should be possible to develop an analytical technique in ecology; and that this approach would enable ecologists to form some fundamental concepts with respect to ecosystems and to purify their language.

Take, for example, the very important term "competition," which was the subject of a fairly exhaustive review by Milne (1962) as part of a

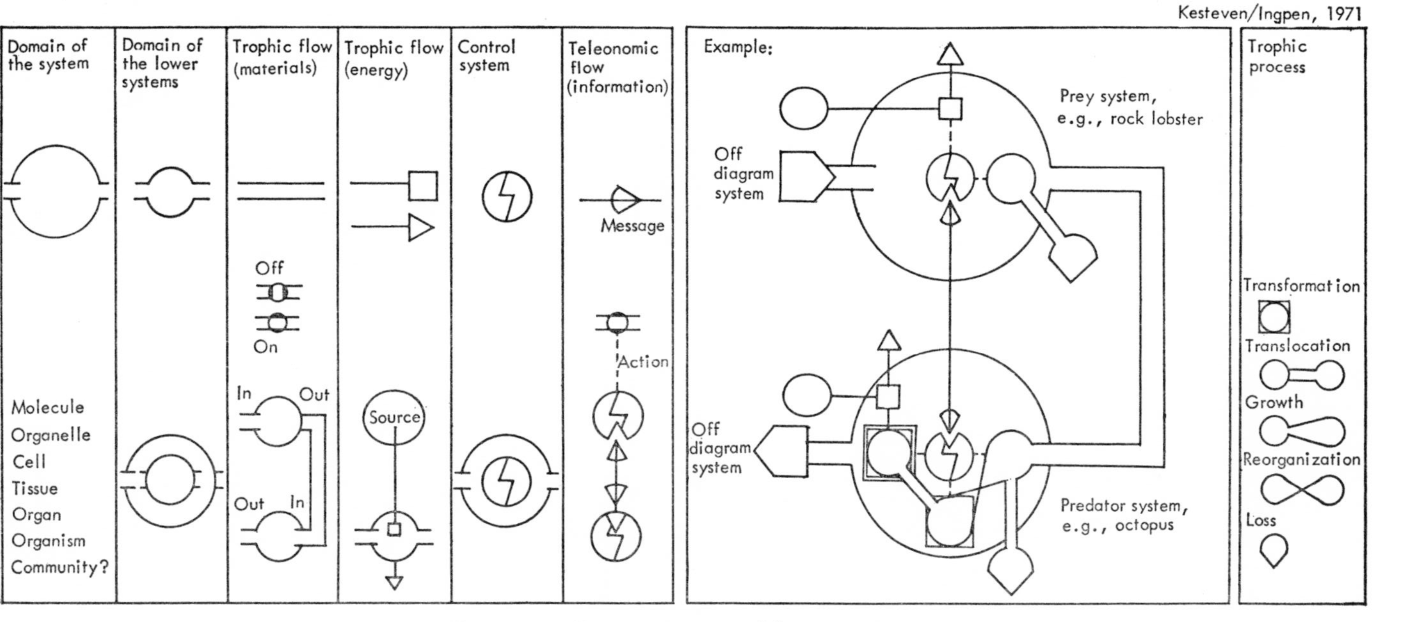

Figure 11. Representation of biotic systems

symposium on mechanisms of competition held by the Society for Experimental Biology. Milne examined a large number of uses of the term, ranging from "direct physical conflict" to "the totality of the consequences of occupying a single *lebensraum!*" These definitions say that competition is (1) a process, (2) an interaction, (3) a demand, (4) the endeavor (Milne's own definition), (5) a situation, (6) what arises, (7) what occurs.

The first two clearly signify an intention to refer to an action taking place. Numbers 3 and 4 seem in a sense to point to motivation, or even to attitude of mind. Number 5 is noncommittal, signifies a state of affairs, while 6 and 7 may refer to "an action that is consequence of" or to "the consequence of action taken." If, however, in obedience to Birch's own precept (1957), we seek to become clear in our minds as to what it is of which we think when we use the term "competition," and if, for this purpose, we seek to identify the fundamental relations represented in Figure 11 (in situations that have had the characteristic that lead us to employ the term "competition"), we immediately find that what we are thinking of is not a relation, nor even a process, of itself, but a quality attributed by us, sometimes to relation, sometimes to Processes II, and sometimes to Process I.

That is to say what is in our minds when we utter the word "competition" is a concept neither of something structural, nor of something functional, nor of any event, but rather an apprehension of the significance of a whole complex of (1) passive and active relations, (2) actions taking place through those relations, and (3) consequences of those actions. Moreover, it is clear from the literature that authors have in mind not only an immediate consequence, such as the death of one animal in a struggle for possession of some object, but also the long range consequences genetically of the elimination of the genetic potential of animals so defeated.

I know of no single term with which to refer in general to such a total complex, let alone to the significance of such complexes. I cannot point to any instance of such a complex saying, "That is an *X*," and by the word "*X*" convey all I want to convey. Words such as "situation," and phrases such as "state of affairs," are ineffective for this purpose. Nor, naturally, do I know of any construction of the form, "This is an *X* of the type *A*," by which I may also convey a correct assessment of significance. It is of little help to refer to a *situation* of symbiosis or of parasitism or of commensalism or of predatism, for each is little more than a mandala to evoke from each reader's mind whatever he has learned about such complexes *in general.* Yet it is to the significance of whole complexes such as these that we must direct our attention in ecology. Furthermore, it is my argument that we must identify these complexes and then strip them down to their fundamental relations and processes so as to be able to reconstruct them in a meaningful way.

The word "game," in the theory of human behavior, belongs to the class of terms to which the term I seek will belong, but of course it does not serve my purpose. However, the word "game" will serve to illustrate further the problem as I see it. We can describe the actions of a game, such as golf, in the simplest of physical terms, without reference to rules, aesthetic responses, or other features of the game, thus making the activity appear entirely ludicrous; or we can describe the game in terms of the professional golfer's deep understanding of the manner in which, by choice of club, use of information about the course and about contemporaneous weather, and muscular and nervous control, he seeks to make each stroke achieve a best result; and we can go on to include in our description some reference to the total pattern of the game, to the motives of those who play it, the enjoyment got from it and so forth; and in the end we may achieve a full account of this particular complex and an assessment of its social significance.

Thus "game" is the general term, and "golf" and "football" and others are particular instances. In ecology we lack the general term, and lacking even the concept of it, we have not yet established a procedure for formulating comprehensive accounts of its instances. In that situation, our use of terms like "competition" is like labeling some game as "soccer" without knowing the shape of the ball or the rules being observed, and in some cases is like calling some activity a game of rugby union when it might instead be a riot.

The problems confronting us are represented also in our terms relating to structural elements. In the language of Andrewartha and Birch (1954) the environment of an animal is "all those things or qualities which influence its chance to survive and reproduce"; and here again we have a prior attribution of significance by selection from among the characteristics of that of which the authors are writing. Moreover, as Milne (1962) points out, the scheme of environmental analysis supporting this definition is itself debatable. I suggest that we must see this term "environment" as an element in the array: (1) site, (2) habitat, (3) environment, and (4) niche.

Let us suppose the first to denote an abiotic and completely neutral entity. An instance of it would be designated by geographical coordinates; upon study it would reveal physical and chemical characteristics. Let us next suppose the second to denote the place where a plant or animal lives and all the characteristics of such a plane. Of course a habitat must have location, and therefore basically it is a site, but within that site there are other organisms, and the habitat characteristics must therefore include features resulting from the presence of other biota; moreover it will be obvious, from evidence relating to a great many species, that in its life an organism may occupy a number of habitats. If precise connotations of the terms "site" and "habitat" are thus assumed, "environment," unless it is a synonym of one or other of the first two, can denote only that set of influences, from among all the influences emanating from the components of its habitat, that influence an organism in any way; and environment must be unique to each organism. These influences, in interaction with the characteristics of the organism itself, must determine everything the organism does, and it can be only a very special view that sees these influences only in respect to their determination of the probability that the organism will survive and will reproduce.

The word "niche" has come to have a most curious penumbra of confusion. Weatherley (1963) chooses to regard it as signifying only an organism's trophic relations, and of course if there should be common agreement so to restrict the word, it could serve in this fashion. However, in the original usage by Elton, the term signified of an animal "what it does," the role it plays, and in this sense "niche" is, of an animal, that set of relations (in complement to those that constitute its environment) that have their origin in the animal itself, and which in interaction with the characteristics of the habitat determine the biotic role and significance of the organism.

Finally, we can illustrate these problems by an examination of classificatory hierarchies with special reference to the word "population." In Figure 12, we have represented a biotic hierarchy from molecule to community. The essential characteristic of this hierarchy is that an individual of one level is, by virtue of the integrative consequence of its control system, a complex of characteristics which is more than a statistical sum of the properties of the systems of the lower level of which it is composed. However, we have to remember that each word signifying a level of the hierarchy in the diagram is only the name of the level and that at each level we have in one direction a sequence: individual, numbers, aggregations, populations; and in another direction we have taxonomic sequence, such as variety, subspecies, species, genus, family, etc. In this sense we have a three-dimensional conceptual matrix. On one coordinate is the sequence of levels of organization of individuals; on another is the number sequence; and on the third is a genetic, or at least a taxonomic sequence. The special point I want to make at this stage is that the word "population" belongs on the second sequence and has validity both at any level of the organizational hierarchy and at any level of the genetic hierarchy. That is to say, we can have a population of cells, equally well as we have a population of organisms; similarly, at the organismal level we can have a population of a species or of any other taxon; and at community level we can have a population of a biocoenose, comprising members of a diversity of taxa. In consequence, the current tendency to write, say, of "populations ecology," as though population were itself an organizational level separate from organism or community, is erroneous. Of course, it is obviously meant to denote what was intended (and by some is still intended) by the term synecology, but its users would probably be confounded if confronted with a usage of it in connection with a population of cells. Moreover, the term seems to be a retrogressive step in the presence of current thought about a species which is so firmly based in the principle that the individual representative and the population of a species are separable only methodologically.

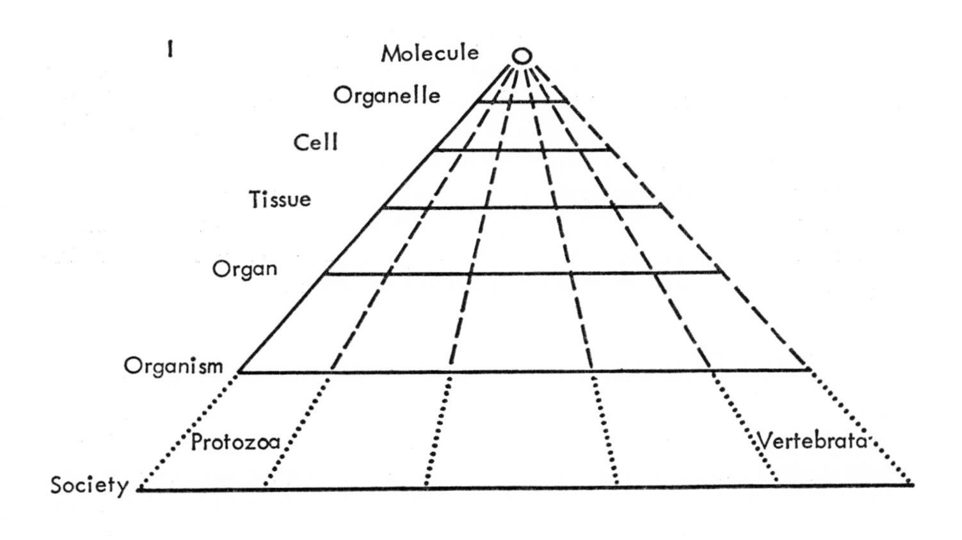

II

Molecule

Organelle

Cell

Tissue

Organ

Organism

Community

Human Society

Figure 12. Alternative schemes for placement of supra-organismic entities in a biotic hierarchy

The Necessary Revolution II: In Ethic

I have argued above against maintenance of the concept of common property resource and in particular against adducing from that characteristic an argument for free competition and minimal control in the exploitation of fishery resources. I have also argued that the future of our resources is being shaped by management whose characteristics, in the future, would be greatly influenced by realistic participation (in management) of each fisherman, each fleet, and each nation, and that that participation would be motivated by a self-interest which recognized the relation of that interest to community interest. I have further argued that such self-interest would be satisfied, and achieve confidence, only by analysis in depth of the natural, economic, and social systems involved in each fishery, by the accuracy of description of those systems, and by the reliability of predictions as to their behavior under various conditions. That confidence would make itself felt in a technical plan to which each operative would be party and of whose implications he would be well informed, but further I argued that in turn these conditions of confidence implied a political arrangement as to the allocation of participation in the exploitation of a resource. What I want to discuss now is the nature of such a political arrangement and the logical basis on which it would be formulated, with due regard, however, for a number of nonscientific elements which of necessity enter into any political arrangement.

Some of my earlier remarks concerning the nature of natural systems and of the biotic hierarchy have a relation with the bases of what I call political arrangement. We can suppose the biotic hierarchy, which perhaps begins with organic molecule, to lie through organelles, nuclei, cells, tissues, and organs to organisms. It seems only reasonable to suggest that above organism lies a supra-organismic level in direct continuity with the rest of the biotic hierarchy. The individuals of this level, if it existed, would be distinguished from those of the immediately lower level (as each other level is distinguished from what lies below it) by characteristics which are more than the sum or a statistical synoptic of characteristics of individuals assembled at that lower level. Organisms being assembled as stock, population, or community, it is to be asked whether these constitute entities at supra-organismic level, in the biotic hierarchy, and in respect of this my thought is as follows. It seems to me that we can show that that which constitutes and establishes entities at one level, securing their persistence as entity, directing their reproduction and hence their projection through time, is that some part (or parts) of the assemblage of individuals of the lower level is (or are) specialized and in some way set apart to constitute a center or system of control; the behavior, in every sense, of the entity is dominated by the operation of that center or system. At organismic level the role of brain, or other nerve center and system, and of various humoral systems, controlling or directing the orchestra of which the individual is composed, is plain enough. And at cellular level the role of the nucleus also seems plain. But, when we come to assemblages of organisms, what kind of setting apart and specialization can we recognize? Is there, for example, some designation of individuals in a flock of birds or a school of fish, and specialized modification of them, to integrate and direct that assemblage? Is there such a center or system in a natural community? I believe that there is not and that we cannot look for anything corresponding physically to brain or nervous system at supra-organismic level. A population is to me merely an assemblage and the word population, validly applicable at all levels of the hierarchy, is merely a way of signifying some special assemblage of individuals at each level; thus we can speak of a population of cells equally with a population of organisms. A natural community exists as such virtually by accident; its individuals relate to one another ecotically subject to the physical and chemical circumstances of its habitat; the bounds of the community are accidentally set by topographic and other physical and chemical characteristics, and when those characteristics of habitat change, the limits of the community change. There is nothing in a community either to determine its behavior or to give it persistence or to secure its recurrence. If a community recurs it is not because of what might have happened in or been done by any previous instance of that community acting as an entity. In particular I would say that we cannot conceive of any synod of seals or council of crocodiles which assembles at regular intervals to determine the feeding, reproductive, or other behavior of the individuals of a community of those organisms.

The case of man, however, is different. Man is aware of himself and of his fellows, and his faculty of memory is undoubtedly larger and more extensive, because of his more efficient and more cultivated faculties of perception, than that of any other organism. Moreover, he has somehow acquired the use of symbols by which he is able to communicate his thought and thus to organize and direct not only his own activities with a view to the future but also those of his fellows. In consequence, human society possesses characteristics differing from the characteristics of assemblages of any other species. Undoubtedly the communities of some other organisms manifest features which have the appearance of being tentatives toward the characteristics exhibited by human society. Thus among some animals we see care for the young; we see an intensive form of community organization in various insect communities. But in every case these activities are carried out in fixed patterns, varied only in relatively few cases and to a relatively small degree; in no sense do these modifications correspond to the inventiveness of man and the manner in which he can deal with contingency.

A friend of mine believes that he summarizes this situation when he says: "Con la vida el universo empezó a verse, escucharse, sentirse y con el hombre a pensarse." That is: with life the universe began to see itself, hear itself, feel itself, and with man it began to think about itself. Anyone who can go with me so far in my argument can probably continue with me and agree that man, with this awareness of himself, his fellows, and his universe, with his symbols, with his power to organize, and with his power to modify his habitat, carries responsibilities not only to himself but to the whole of his universe. If one agrees with that, then he will be willing at least to examine the following matters with me.

First, I wish to look at the matter of the use of natural resources. Man, of his nature, takes from his habitat that which he needs for survival as individual, generation, and race; but, more than this, he takes these resources for the totality of his physical, mental, and emotional experience. Yet at the same time he is obliged to recognize that there are limits to these resources. Obviously, while each nonrenewable resource exists in only a fixed quantity, the frequency of re-use of virtually nondestructible resources is without limit; among these, perhaps the most notable and most important is water. Thus while there are some resources which apparently we can use up, and which we shall not be able to have again, there are others which we can use and re-use; but in the second case the total availability at any point in time is fixed. We may change the rate of cycling but even so as we move from one level of technology to another we are confronted at each with absolute limits to availability at one point of time. Next, there are renewable resources which, obviously, need not be destroyed but, equally obviously, can be. We have much evidence to show that we can eliminate a particular species. Moreover, each species or variety is limited at any particular time as to the population it can achieve: with our technologies we can modify these limits, we can modify the patterns in which populations relate to one another; we can modify the capacities of particular species and varieties to make them serve particular purposes we have for them; but there is always some absolute limit. The choices of mankind, as a whole, are innumerable, so long as we are talking about all the choices offered to each man separately, to each group and to each nation. But, as the world population grows, the possibility of taking a decision without affecting other people becomes less and less.

To me this course of events implies that we need to examine three major principles. The first of these is the principle, of most religions, that "I am my brother's keeper": I have a responsibility for his well-being and I cannot, with clear conscience, ever act without thinking about the consequence to other people of what I do. As a corollary to this we must reject fiercely, I believe, all arguments of the sense that it is good to buy cheap and sell dear. To me it seems that this is vicious, as is every other deliberate exploitation of one's fellow men, and of all vices it is among the most dangerous. He who buys cheap denies due reward to the person from whom he bought and steals from the person to whom he sells; such a person sees only virtue in free competition and supposes the "law of supply and demand" to be current legislation. The failure to observe this principle and its corollaries is a major part of the reason why we fail to close the gap between the underdeveloped and developed countries.

The next major principle which I believe must be examined is that of the need to return to parsimony. The exercise of parsimony in simple economic situations in the past was the necessary path to security. Of course security could not be guaranteed by this path but at least this path was the best insurance against natural calamity. The necessity for parsimony arose partly from the fact that such communities were isolated and had no means of appealing in times of scarcity to other communities for help. I believe it is no exaggeration to say that we are rapidly approaching a point of world development at which we, as the total human race, must recognize that our situation on this world corresponds to that which confronted an isolated farmer in an isolated community of centuries ago: there is no other community or resource to which we can appeal in times of necessity.

This leads to the third of my major principles, namely, the need for the exercise of humility. Parsimony is itself a humility: its contrary is the arrogant taking from other people. No one is entitled to an immensely rich and luxurious way of life at the expense of others, and, *a fortiori*, to have these privileges cannot be justified merely by possession of know-how, nor because historically one is fortunate in one's social institutions. But here I must express a view about the term "historical force": I believe there is no such thing. I hold that no event of the past reaches forward to exercise physical or psychological effect on the present. Every act of today is the result of forces now and every decision of today is a decision of today. No order issued in the past has any effect today except by virtue of decisions made today in accordance with the totality of information available to us now. You may ask me, what about the immense volume of law which guides our courts and our administrations, and I reply that the word "guide" indicates precisely what I am saying. A writ runs by force or consent: physical force can be exerted only by the living and only the living can give or withhold consent;

and courts daily revise the decisions of previous courts. That is to say, what has come to us from the past takes, in each moment, a new existence as time flows past, and everything is subject to some change. Every physical object is different in some degree, however slight, from one moment to the next and every transmitted thought is different from one moment to the next. Each physical object changes because of forces within itself and of forces about it, but all these forces are physical. The significance of a thought changes because of what is going on in its intellectual environment, which is, I think, what Pierre Teilhard de Chardin called "noosphere" and in my view each unit of thought is as much subject to change internally and because of external effects upon it as is each physical object. We can see a simple example of this in the semantic changes which particular words have undergone over centuries. My point then is that man with his capacities and responsibilities is himself a changing thing, standing within a vast changing complex, and our ethic must be to find ways of managing ourselves individually and collectively in this situation. Each of us has his own particular role to play. The principal concern here is with the role of scientists and, among scientists, with the role of ecologists. I have for long thought that one particular feature of the changes taking place is that ecology which is now becoming a field of major interest and preoccupation, is only John the Baptist to the Messiah of intellectuality—the social sciences—for I agree with Pope that the proper study of man is man. An ecologist studies the simpler relationships in which organisms stand to one another and the simple relations between them, which include some of the relations which man has with his habitat. But, after all, what is studied by ecologists is only the more elementary aspect of that totality of relationship in which man stands to his universe and our urgent necessity is for a better understanding of that total relationship, more especially of the relations that prevail between individual men, between groups of men, and between nations.

April 1971

REFERENCES

Andrewartha, H. G., and L. C. Birch. 1954. *Distribution and Abundance of Animals.* Chicago: University of Chicago Press.

Birch, L. C. 1957. "The Meanings of Competition." *American Naturalist,* 91 (No. 856): 5–18.

Bronowski, J. 1971. *The Identity of Man.* Rev. ed. Garden City, N.Y.: Natural History Press.

Glass, B. 1956. "The Scientists' Perspective." *Science,* 123:777.

Ivlev, V. S. 1961. *Experimental Ecology of the Feeding of Fishes.* New Haven, Conn.: Yale University Press.

Kesteven, G. L., and R. R. Ingpen. 1966. "Representation of the Structure of Biotic Systems." *Australian Journal of Science,* 29 (No. 4): 97–102.

Kuhn, T. S. 1962. *The Structure of Scientific Revolutions.* Chicago: University of Chicago Press.

Milne, A. 1962. "Definition of Competition among Animals." *Proceedings of the Society for Experimental Biology,* 15:40–61.

Needham, J. 1954–70. *Science and Civilization in China.* 4 vols. Cambridge: Cambridge University Press.

Weatherley, A. H. 1963. "Notions of Niche and Competition among Animals, with Special Reference to Freshwater Fish." *Nature,* 197:14–17.

Index